Lecture Notes in Economics and Mathematical Systems

416

W0245752

Lecture Notes in Economics
and Mathematical Systems   416

Herman S. J. Cesar

Control and Game Models of the Greenhouse Effect

Economics Essays on the Comedy and Tragedy of the Commons

Springer-Verlag

Berlin Heidelberg New York
London Paris Tokyo
Hong Kong Barcelona
Budapest

Author

Dr. H. S. J. Cesar
Tilburg University
Department of Economics
Room 11-22
P. O. Box 90153
NL-5000 LE Tilburg, The Netherlands

ISBN-13: 978-3-540-58220-5   e-ISBN-13: 978-3-642-45738-8
DOI: 10.1007/978-3-642-45738-8

CIP data applied for.

Typesetting: Camera ready by author
SPIN: 10474201   42/3140-543210 - Printed on acid-free paper

"O sol che sani ogne vista turbata,
tu mi contenti sí quando tu solvi,
che, non men che saver, dubbiar m'aggrata.

Ancora in dietro un poco ti rivolvi",
diss'io, " là dove di' ch'usura offende
la divina bontade, e 'l groppo solvi".

"Filosofia" mi disse, "a chi la 'ntende,
nota, non pure in una sola parte,
come natura lo suo corso prende

dal divino 'ntelletto e da sua arte,
e se tu ben la tua Fisica note,
tu troverai, non dopo molte carte,

che l'arte vostra quella, quanto pote,
segue, come 'l maestro fa 'l discente;
sí che vostr' arte a Dio quasi è nepote.

Da queste due, se tu ti rechi a mente
lo Genesí dal principio, convene
prender sua vita e avanzar la gente;

e perché l'usuriere altra via tene,
per sé natura e per la sua seguace
dispregia, poi ch' in altro pon la spene ...."

Dante Alighieri, La Divina Commedia, Inferno, Canto XI, 91-111
(a cura di Pasquini/Quaglio, Garzanti, Milano, 1990).

Preface

This book is the result of a four years' research project at the European University Institute in Florence, Italy. I am grateful to my supervisor, Mark Salmon, for his many ideas. I am also indebted to my co-supervisor, Aart de Zeeuw, for his unfailing support and to Carlo Carraro, Louis Phlips and Alister Ulph.

I gratefully acknowledge the help and assistance of many people that advised me at some stage, Scott Barrett, Peter Bohm, Lans Bovenberg, Partha Dasgupta, Klaus Hasselman, Peter Kort, Olli Tahvonen, Cees Withagen and Clifford Wymer.

Thanks also to the faculty and students of the Department of Economics of the European University Institute and especially to Jacqueline, Jessica, Marcia and Barbara for their moral support; my Italian teacher Camilla; my 'Gruppo di Lavoro' Tilman, Peter and Luisa; my dear Dutch friends Yvo, Pieter, Ida and Ellen; international friends Bogdan, Stuart, Melanie, Henning, Anantha, Paolo, Pompeo, Nikos, Christian, Analisa, Dorothea, Valeria and Stefania; the members of the Working Group of Environmental Studies; and finally my girlfriends who gave up on me because they thought I was working too hard and not spending enough time with them. Also many thanks to my landlady Olga, who taught me the beauty of Italian literature and opera and my landlord Emilio and their dog Igor.

I am also greatly indebted to Professor Larry Susskind and to the Massachusetts Institute of Technology where I spent an extremely inspiring semester. Also many thanks to all those who helped me during that stay, particularly Bill Clark, Oliver Fratzscher, David Laws, Ted Parson and David Victor as well as to the participants of the Harvard Seminar on International Environmental Institutions.

Many thanks to Mario Nuti and Timothy King for arranging a very interesting internship and a smashing summer at the World Bank where I came for the first time in contact with the practice of environmental problems and policy.

Also, I am grateful to my former collegues at the Agricultural University of Wageningen, especially Henk Folmer and Ekko van Ierland, who were the first to arouse my interest in environmental economics.

Moreover, I would like to thank the Economics Department at Tilburg University and especially Aart de Zeeuw for giving me the opportunity to finish this work and the CentER for showing me how a research institute can be run efficiently.

Finally I would like to thank my father, mother, brother and sister and all my friends back home and all over the world for their continuous support.

Contents

Chapter 1

Introduction and Conclusions

Nel mezzo del cammin di nostra vita
mi ritrovai per una selva oscura,
ché la diritta via era smarrita.

Ahi quanto a dir qual era è cosa dura
esta selva selvaggia e aspra e forte
che nel pensier rinova la paura!

Tant' è amara che poco è piú morte;
ma per trattar del ben ch'i' vi trovai,
dirò del l'altre cose ch'i' v'ho scorte.[1]

Dante Alighieri (1309).

Life on Earth would be impossible without the Natural **Greenhouse Effect**. This effect results from the capacity of various atmospheric trace gases to absorb infra-red radiation. These gases act as a blanket which helps to trap heat.

Mankind is adding Greenhouse gases, especially carbon dioxide, to the air thereby changing the natural radiative balance. At the same time, an area of forest equal to the whole surface of Italy (that is eight times the size of Holland) is being cut annually, thereby destroying one of the most valuable (carbon) buffers.

---

[1]La Divina Commedia, Inferno, Canto I, 1-9; English translation (D.L. Sayers): *Midway this way of life we're bound upon, \\ I woke to find myself in a dark wood, \\ Where the right road was wholly lost and gone. \\ Ay, me! how hard to speak of it — that rude \\ And rough and stubborn forest! the mere breath \\ Of memory stirs the old fear in the blood \\ It is so bitter, it goes nigh to death; \\ Yet there, I gained such good, that, to convey \\ The tale, I'll write what else I found therewith.*

The effects of this *enhanced* Greenhouse Effect are highly uncertain. Especially uncertain are the ultimate magnitude and the timing of warming. Also highly uncertain are the implications of this warming for the Earth's climate system, the global sea level, the stability of ecosystems and eventually the economy[2].

The major source of Greenhouse gas emissions is energy production and use (fossil fuels) in power-plants, industry and transport.

The aim of **economic policy-making** in the case of the Greenhouse Effect, is to design an efficient mitigation strategy that optimises overall welfare.

In the decision making process, the costs and benefits of reducing Greenhouse gas emissions (and of stopping deforestation) have a central place. The costs include switching from fossil to non-fossil fuels, reducing energy consumption (insulation, etc.), changing combustion technology and so forth.

Amongst the corresponding benefits are the prevention of damages to crops, environmental amenities, ecosystems, etc. as well as expenses saved on strengthening of dykes.

Efficiency requires that marginal costs of mitigation are equalized over all possible forms of emission reductions. This can be achieved with the help of market-based policy instruments.

Control theory describes methods of finding the optimal levels of the instrument or control variables at each instant of time satisfying an intertemporal objective function. An essential feature of these problems are the dynamic constraints through which the controls influence the trajectories of state variables. Examples of control variables are the level of fossil fuel use, the investment in abatement technology and taxes. State variables could be, for instance, the atmospheric concentration of Greenhouse gases and the stock of energy-related capital. Typically, the results can be formulated in terms of the shadow values of the state variables. For example, in the case of pollution, the shadow price represents the value of an infinitesimal deterioration of environmental quality. This value corresponds to the optimal intertemporal emission charge.

In case there are two or more policy makers who influence each others' objective functionals either directly or through the state variables, then strategic interaction becomes important and can be analysed using dynamic **game theory**.

---

[2]EPA (1990; p.8)

A quarter of a century ago, Garrett Hardin wrote an article with the title "**The Tragedy of the Commons**". In this paper, the 'commons' refer to the medieval common pasture, open to all for grazing cattle. The 'tragedy' describes the myopic logic of each individual pursuing his own advantage by increasing the size of his own herd, without taking into account the effect of this strategy on overgrazing[3].

Like Hardin's pasture, the Earth's atmosphere is a common-property resource, endangered by over-use: economic agents do not recognise and hence do not include the build-up of Greenhouse gases in their decision of how much energy to use and therefore tend to use too much. Similarly, Brazilian timber-cutters do not apparently account for the effects of their logging on Climate Change. Although this tragedy has existed throughout history, it seems that the scale of the consequences of such actions may now have become a threat for mankind.

The logic behind the 'tragedy of the commons' is the same as that of non-cooperative behaviour in the famous prisoners' dilemma.

The Florentine writer Dante Alighieri wrote, nearly 700 years ago, a book that he gave the name "**La Commedia**". The epithet 'Divina' was added by later admirers. Dante's masterwork is an allegory of the way to salvation. In the first part Dante comes across three beasts, each symbolizing a sin. The Leopard is the self-indulgent sin (Lust; Incontinence). The Lion is the violent sin (Pride; Bestiality). The She-wolf is the malicious sin (Avarice; Fraud).

The strength of Dante' story is that it offers a vision of Hell, a vision that is remedial; therefore it is the drama of each person's individual choice.

The over-use of environmental resources can be seen as the consequence of such a choice. This implies that the logic of Hardin's tragedy is not inexorable; non-cooperative behaviour in the prisoners' dilemma is not inescapable.

The challenge of managing international commons, such as the Earth's atmosphere, is exactly to avoid its over-use, for instance through information, persuasion, the use of correct prices and international cooperation.

In later chapters, this challenge of avoiding the tragedy of aggravating the Greenhouse Effect will be discussed.

---

[3]Dasgupta (1982) criticizes this idea of Hardin (1968) from an economic perspective. In the first place, Hardin's idea that each person has an advantage of increasing his own herd 'without limit' presupposes that private costs of adding cattle are zero, which is usually not the case in practice. Secondly, depletion *per se* need not be economically sub-optimal. See also Parson & Zeckhauser (1992). In my opinion, Dasgupta is absolutely right and his arguments do make sense in the case of fishery and forestry management. However, with international environmental resources, that have a life-support system, Hardin's biological and Dasgupta's economic perspective coincide.

In **Chapter 2**, the groundwork is laid for the analysis later on. First, the most relevant aspects of the Greenhouse Effect are discussed. The causes, trends, impacts and especially the policy options are highlighted. This elaboration will justify the choice of carbon dioxide emissions (CO_2) as the primary Greenhouse gas in later chapters.

Next, the literature on environmental resource economics using optimal control models is critically surveyed. The decentralised version of the Central Planner models is discussed in order to see whether control outcomes can be sustained by a market economy. This also highlights the instruments of environmental policy available in decentralised economies.

Five elements determining different types of control models on pollution are discussed[4]:

Sustainability Different assimilation functions are discussed in the literature, varying from linear specifications, to highly non-linear ones and to functions incorporating 'flip-flops'.

Pollution Models are classified into 'polluting input models' and 'polluting output models'. In the latter case, pollutants are a by-product of production. Polluting input models, on the other hand, describe pollutants as intermediate goods. An example is the use of pesticides in agriculture, where the crop-production itself is not polluting, but the pesticides are[5].

Emission reduction A decrease of polluting emissions (or discharges) is possible in various ways, ranging from the lowering of production, abatement, recycling and process-integrated changes. Note that the type of pollution critically determines the scope of emission reduction possibilities: water pollution allows for purification after discharge whereas abatement of air pollution is only possible in the polluting plant. This has a critical impact on how 'net-pollution' is modelled, an issue that tends to be forgotten by model-makers without knowledge of environmental sciences.

Capital With a fixed capital stock, the decision-maker can allocate output between consumption and abatement (or recycling, etc.). With capital accumulation, a Ramsey problem with pollution is created. In this case, the decision-maker

---

[4]Models where pollution is modeled as a flow are not surveyed.

[5]Note, however, that 'polluting input models' can be seen as a joint production representation of a 'polluting output model'.

can allocate output to consumption, abatement and net capital accumulation. Human capital aspects are discussed as well.

Gaming aspects Control models with more than one planner (e.g. an international context) form a differential game. The type of pollutant under consideration dictates the way in which the accumulation of pollution is modeled (compare the Greenhouse Effect with asymmetric pollution problems like acid rain).

The existing literature is surveyed focussing on identifying which elements are lacking in current Greenhouse models and which elements need to be added. The aim is to gather the elements that will allow for the development of intertemporal optimisation models of the Greenhouse Effect, appropriate to address the policy questions in the subsequent chapters.

In **Chapter 3**, one-country models of the Greenhouse Effect are developed and four elements, often neglected in the literature are elaborated in particular.

First, a rudimentary model with one state variable (the atmospheric concentration of Greenhouse gases) and without abatement possibilities is worked out. Specific functional forms are taken in order to analyse the features of this model. The build-up of Greenhouse gases is assumed to have negative impacts on productive capacity and not on the people's well-being directly. This is a deliberate choice away from the 'luxury' characterisation of environmental quality. This specification implies that the fundamental intertemporal trade-off is between current consumption and future consumption rather than between current consumption and future amenity. Conditions are derived under which higher time-preference leads to both lower future consumption and more pollution[6].

Secondly, in order to discuss the concept of sustainability, different non-linear specifications of the natural assimilation function are introduced. There exists large uncertainty with respect to the actual assimilative capacity of nature for elevated levels of the concentration of Greenhouse gases. Given this uncertainty, the robustness of policy conclusions to changes in the specification of the natural regeneration function is analysed. The conclusion is that slight variations in the parameters of the assimilation function can have a dramatic impact on the steady

---

[6]In models where pollution has only an amenity effect, a higher time-preference typically leads to higher future consumption and more pollution. Crucial for this result is an assimilation function that is increasing in the level of pollution.

state results. This indicates that explicit treatment of uncertainty in future research is extremely important.

Thirdly, capital accumulation in the energy sector is introduced. This allows for an analysis of the optimal level of energy-related investment versus current consumption and fossil fuel use. The standard stability and comparative static analysis[7] is extended by looking at the trajectories of the state and control variables. These show a considerable overshooting of energy related capital. This means that it is optimal for an economy to start with a rapid build-up of energy technology that depreciates later on.

Fourthly, the human and physical elements of capital accumulation are explicitly modeled[8]. Modelling of human capital is particularly important if externalities in its accumulation are present. The consequences of such knowledge spill-overs are discussed. The externalities are modelled such that production has decreasing returns to scale in the absence of government intervention, however with internalisation, there are constant returns to scale. This is an important first step to modelling endogenous growth in an economy with pollution, where sustained growth depends on economic policy. In the decentralised version of the model, the externalities in both pollution and knowledge are internalised through carbon taxes and human capital subsidies respectively.

In **Chapter 4**, the issue of the 'tragedy of the commons' is highlighted by looking at the transboundary aspect of the Greenhouse Effect. To clarify this, assume the following prisoner's dilemma game of a world consisting of two countries. The economic activities in both countries are largely based on fossil fuel use. Shifts to renewable sources of energy are available, but more expensive. Assume that a country can either continue with fossil fuel use, the net benefits of which are normalised to zero or can shift to renewables. The costs of this shift are 6 ECU and the benefits (in terms of a reduced danger of Global Warming) are 5. If both countries substitute away from fossil fuels, the corresponding benefits for each are

---

[7]Conditions under which the steady state is locally asymptotically stable are analysed. Comparative statics results for changes in the time preference are presented. For the numerical specification of the model, it is shown that more patience means in the long run a higher level of consumption, abatement capital and environmental quality.

[8]In a general specification of the models, the conditions are analysed for locally asymptotical stability of the restricted dynamic system where the economy is in a pollution steady state. Trajectories for human and physical capital are discussed for different starting values of the state variables. The results are in line with standard two-sector growth models.

taken to be $5 + 5 = 10$. These assumptions are summarised in Table 1.1.

		Country II	
		fossil	renewable
Country I	fossil	$(0,0)$	$(5,-1)$
	renewable	$(-1,5)$	$(4,4)$

Table 1.1: *Pay-off matrix of a game of fossil fuel versus renewables*

In this matrix the tuple $(5,-1)$ means, for instance, that country II shifts to renewables, bearing costs of 6 and earning benefits of 5, so that its pay-off is $5 - 6 = -1$. Country I has the same benefits of reduced danger of Global Warming but at no costs, so that its pay-off is $5 - 0 = 5$. The other pay-off tuples can similarly be calculated.

The tragedy is that each country individually prefers to stick to fossil fuel use, whatever the preference of the other country, but both are better off by cooperating and shifting to renewable energy. This problem is fundamental to the suboptimal provision of public goods $((0 + 0) < (4 + 4))$. The Greenhouse Effect has been labelled the 'granddaddy of all public goods by Nordhaus (1991) as it concerns all people in the world now unto the indefinite future. In economists' parlance, the spill-over effects form an externality that is not appropriately valued. The resulting inefficiency can, in principle, be corrected by international cooperation aimed at internalising this externality. The theory of public economics suggests that appropriate taxation and lump sum redistribution can lead to first-best results in the presence of externalities. The design of such policies is the subject matter of this thesis.

Although the example above highlights some of the strategic issues raised by the international nature of the Greenhouse Effect, it is too stylized. Instead of discrete (two) policy options and one time period, reality demands we consider continuous dynamic choice problems and hence in Chapter 4 an optimal control framework is used. The state variables are taken to be the atmospheric concentration of Greenhouse gases and the stock of energy-related capital. The controls are effective fossil fuel use and capital investment.

In a multi-country setting, this framework creates a differential game[9]. The

---

[9]Chapter 4 will start with a review of the differential games.

introduction of energy-related capital in such a game is quite novel[10]. This will widen the policy options available to a single government in choosing an appropriate policy response and presumably make the analysis more realistic and relevant.

The conclusion of numerical calculations is that cooperation leads to substantially lower levels of fossil fuel use and hence of the atmospheric concentration of GHGs. The effects of concerted action on technology is quite marginal, the impact depending on substitution possibilities between fossil fuel and energy technology.

Furthermore, social welfare appears to be lower in the feedback case than in the open-loop solution. This means that potential gains from cooperation are even more elevated with the feedback solution instead of the open-loop equilibrium as a benchmark case[11].

The clear benefit from concerted action raises the question as to why cooperation is difficult to achieve in real life. The prisoner's dilemma described above explains this tragedy and raises the challenge of how to avoid the free rider problem, and how to preserve cooperation with self-enforcing agreements in differential games.

In fact, as the Folk Theorem states, cooperation can be sustained once the prisoner's dilemma is repeated over time, using trigger strategies[12]. The problem with trigger strategies is that, by punishing, a player hurts not only the reneger but also himself/herself. An alternative are renegotiation-proof strategies[13].

---

[10]Other examples include Xepapadeas (1991) and Van der Ploeg & Ligthart (1993). These models describe pollution games where both the pollution stock and the stock of 'technology' have a public good character. In Chapter 4, it is assumed that energy-related capital does not have public-good aspects.

[11]Note that in a case with uncertainty and with only one actor, the feedback solution dominates the open-loop solution. The reason is that the actor can adapt when the circumstances require this. On the other hand, as discussed above, the feedback solution is worse in a situation without uncertainty but with two actors for our model. The reason is that countries anticipate that emission increases in one country push the optimal level of emissions in the other countries down. Hence, the marginal benefits to the environment of an additional unit of emission reduction are lower in the feedback case than they would be in the open-loop Nash equilibrium. My conjecture is that in a situation with both uncertainty and with two actors, the feedback solution can be better or worse (in welfare terms) than the open-loop solution, depending on the specification of the model.

[12]Trigger strategies refer to the situation in which each country cooperates up to the moment that one of the members of the agreement reneges. After this has happened, the other country implements non-cooperative energy strategies for some period until cooperation returns.

[13]In the example above this corresponds to following strategy: Begin in the cooperative phase where both players play 'renewable'. If a single player i deviates to 'fossil', switch to the punishment phase for i. In this phase, player i plays 'renewable' and the other player plays 'fossil',

Therefore, trigger strategies and renegotiation proof strategies will be introduced for the differential games on pollution. The conclusion is that cooperation can, in principle, be sustained using either trigger strategies or renegotiation proof strategies, as long as cooperation is in the interest of all the negotiating parties involved. The results obtained by this analysis are necessarily abstract and cannot be translated directly to real life negotiating situations. On the other hand the results highlight some of the essential issues, that need to be dealt with in International Environmental Agreements, such as the issue of how to punish cheaters in a self-enforcing way and how to deal with the question of monitoring compliance.

If Canada were, as is alleged, better off in a 'globally warmed' world, why would it have an incentive for joint action to curb emissions of Greenhouse gases? And why would Third World countries with massive underdevelopment and poverty agree to spend money on an issue like Climate Change that may be far beyond the scope of their interest.

These two questions, discussed in **Chapter 5** raise the point of asymmetry between actors that could impede the development of self-enforcing agreements. This can be analysed in the following asymmetric prisoner's dilemma game. Assume the example given in Table 1.1 with the difference that country II now has costs of 5 ECU and benefits of 2. The corresponding pay-offs are summarised in Table 1.2.

		Country II	
		fossil	renewable
Country I	fossil	$(0,0)$	$(5,-3)$
	renewable	$(-1,2)$	$(4,-1)$

Table 1.2: *Pay-off matrix of a game of asymmetric fossil fuel versus renewables.*

Note that the cooperative solution still gives the socially optimum result $(4 + (-1) = 3)$. However, country II is now worse off compared to the non-cooperative Nash equilibrium. Therefore, cooperation cannot be sustained any more when this game is infinitely repeated.

In such situations side-payments could save the cooperative solution. Chapter

---

after which both players restart cooperation. Note that now, punishing is in the interest of the punisher. A proper definition of renegotiation proofness will be given in Chapter 4.

5 considers that there are incentives raised by such lump-sum (pecuniary) redistribution that impede potentially beneficial (first-best) cooperation. Two ways of re-establishing some form of joint action are elaborated, giving rise to second-best types of cooperation to curb the Greenhouse Effect. These are:

- donating emission-reduction devices in the form of **technology transfers**;

- combining two roughly offsetting problems by means of **issue linkage**.

The first type of cooperation, technology transfers, is stressed forcefully by international organisations (e.g. OECD (1990), UNCED (1991)[14]). In this case, an efficient global emission-reduction can be brought about by investments of, for instance, the 'North' in the 'South'. The crucial merit of technology transfers is that the 'South' can increase its energy efficiency quite dramatically with relatively low costs for the 'North'. Apart from the positive impact on the Greenhouse Effect, this is beneficial for the 'South' as it means less waste of expensive (often imported) resources. Additional benefits to the 'North' are the result of the fact that marginal costs of emission reduction are often much lower in the 'South' than they are in the 'North'. Besides, it can stimulate employment at home and at the same time, in-kind transfers give the 'North' the possibility to make sure that the transfers are actually used for emission reduction.

In the example in Table 1.2 above, the idea of technology transfers can be captured by assuming that country I has a technology that it cannot use at home, but that could decrease its neighbour's costs by 2 ECU. The costs of this technology for the donor are 2.5 ECU[15]. If country I gives this technology (in-kind) to country II, this would decrease the pay-off of the donor to $4-2.5 = 1.5$. The recipient would see its pay-off increase to $-1+2 = 1$. Note that the transfers make the cooperative solution individually rational for both countries, so that it can, in principle, be sustained. The assumption that lump-sum side-payments are infeasible means that cooperation with technology transfers forms a second-best optimum.

In Chapter 5, it is shown that (in-kind) technology transfers can overcome some of the incentive problems that render cash transfers prone to strategic behaviour. The issue that lack of commitment of the 'South' to the continuation of current emission-reduction efforts could make technology transfers vulnerable to new forms of strategic behaviour is also dealt with here. In both cases, conditions are worked

---

[14]For an overview on technology transfers, see Stewart (1990)

[15]The difference can be due to additional costs of the transfer, such as transportation, training of local officials, etc.

out under which such transfers of emission-reduction technology lead to higher consumption and lower Greenhouse gas concentration. As an illustration, some simulation results are given.

The second way of re-establishing some form of second-best cooperation between asymmetric actors is the use of **issue linkage**. Issues are said to be linked (or 'added' or combined) when they are simultaneously discussed for joint settlement. A famous example is the linkage of deep seabed mining with unrelated maritime issues in the Law of the Sea Convention that resolved the stalemate situation that existed on each issue separately.

The idea of issue linkage can be explained using the pay-off structure given in Table 1.2. Assume that there exists another issue that is an exact mirror image of the game above. Then, by adding these two games, each country would have four strategies: cooperating in both games, cooperating in one of the games (two times) and not cooperating at all. The total pay-off of cooperation for player I is 3: the sum of his profits from the first game (4) plus the gains of the second one (-1). The same holds for player II.

Folmer, van Mouche & Ragland (1991) were the first to work out the idea of linkage in the environmental literature, looking at static supergames. Cooperation is sustained by the use of trigger strategies. In Chapter 5, their analysis is extended in several ways. First of all, the concept of differentially valued, roughly offsetting issues is defined. The idea is that full cooperation could be sustained in linked games as long as the 'size of the game' and the asymmetries are such that the games are roughly offsetting. This point is proved using trigger strategies as well as renegotiation proof strategies. First of all, this is elaborated in discounted, infinitely repeated two-by-two games (such as the Prisoners' dilemma). Next, the analysis is extended to differential games. Finally, some examples will be hinted at for possible linkages in order to resolve the stalemate situation that exists with respect to the Greenhouse Effect. These are first, the linking of the Global Warming issue with trade/technology issues in the North-South debate and secondly, the linkage between security and global environmental issues between the US and the EEC.

The conclusion is that, in international environmental bargaining situations, linkage may be used much more often than has been done up to now.

Chapter 2

Literature Review

Ed elli a me: "Ritorna a tua scïenza,
che vuol, quanto la cosa è piú perfetta,
piú senta il bene, e cosí la doglienza.

Tutto che questa gente maladetta
in vera perfezion già mai non vada,
di là piú che di qua essere aspetta"[1]

Dante Alighieri (1309).

2.1   Introduction

Nowadays, environmental problems are among the major concerns in the world.
The Greenhouse Effect, ozone depletion and acid rain are high on the agenda
of policy makers. Likewise, these issues have rapidly gained popularity among
economists. The current interest in environmental problems, that gained considerable momentum with the publication of the Brundtland Report (WCED-1987), can
be seen as a second 'hausse' in global environmental consciousness, twenty years after the first big boom of global 'Doomsday' concerns. The two booms differ in that
the current one is more concerned with global environmental pollution whereas the

---

[1]La Divina Commedia, Inferno, Canto VI, 106-111; English translation (D.L. Sayers): *"Go
to," said he, "hast thou forgot thy learning, \\ Which hath it: The more perfect, the more keen,
\\ Whether for pleasure's or for pain's discerning? \\ Though true perfection never can be seen
\\ In these damned souls, they'll be more near complete \\ After the Judgment than they yet
have been."*

first one dealt primarily with the depletion of exhaustile resources[2]. The common
element in both 'hausses' is the concern with natural resources, which, following
Conrad & Clark (1987) consist[3] of:

renewable resources such as terrestrial and aquatic biomass (e.g. forests and
  fish respectively);

exhaustible resources such as minerals, metals and fossil fuels; these are also
  referred to as non-renewable resources;

environmental resources such as soil, water and air quality (or their counter-
  parts soil, water and air pollution).

Basically, renewable resources have a regenerative capacity which makes periodic
harvesting possible. Exhaustible resources, on the other hand, deplete through
harvesting (extraction)[4]. Note that depletion of renewable resources can also occur
due to over-harvesting. This can ultimately lead to the extinction of species.

More often than not, environmental resources have both an exhaustible and a
renewable character. The exhaustible aspect is due to the irreversible characteristic
(at an economic time scale) of many natural resources. In the case of acid rain, for
instance, annual depositions below a certain threshold do not affect the soil acidity
(renewable aspect). However, beyond a critical point, calcium and other buffers
(exhaustible resource) are slowly being depleted irreversibly (see Aalbers, 1993).
The ozone problem, on the other hand, is conveniently analysed in an exhaustible
resource framework, while some water quality issues can be appropriately described
as purely renewable resources.

The definition of sustainable use of natural resources is depends crucially on the
type of resource (exhaustible or renewable), as is elaborated in Section 2.3 of this
Chapter. This should be borne in mind when modelling a specific type of resource.

---

[2]A much earlier 'boom' in concern for natural resources (especially land) began some 200
years ago, when classical economists, primarily Malthus, Ricardo and Mill started to focus on the
scarcity of land and its long-run implications for population growth and per-capita wealth (see
Pearce & Turner, 1990, p.6). Discovery of new lands and mass migration (America, Australia,
etc.) made this issue of land scarcity less acute later on in the nineteenth century.

[3]Instead of Conrad & Clark's terminology, Barbier & Markandya (1989) use respectively the
terms renewable, non-renewable and semi-renewable resources. The terminology of Smith (1977)
is similar to the one of Conrad & Clark, but Smith distinguishes natural and environmental
resources, where natural resources are subdivided in renewable natural resources and exhaustible
natural resources.

[4]On a geological time scale (more than a million years) regeneration is possible, but this is for
the analysis in economic time scale (less than a few hundred years) not relevant.

Environmental resources nearly always have the character of common-property and/or open access resources[5], where, more often than not, property rights for environmental resources are not clearly defined and/or enforcement is not possible. Government intervention is typically needed in order to prevent a 'tragedy' of the commons that stems from the public good character of the resource[6,7]. However, many of the most serious environmental problems (Greenhouse Effect, ozone depletion, acid rain) are "international or even global in scope, and there does not exist a 'supranational government' empowered to intervene in such cases" (Barrett, 1991, p.1) It is exactly the challenges arising due to the international or transboundary nature of problems such as the Greenhouse Effect that will be analysed.

The primary aim of this dissertation is, in the context of the Greenhouse Effect, the economic policy analysis of:
- the determination of the optimal sustainable level of pollution and consumption;
- the way to achieve these levels, especially in an international setting.
In order to tackle these issues, the modelling of the interdependence between the environmental resource and the relevant economic aspects should be dealt with. Of crucial importance for such a model is that it captures the role of the *price mechanism* and of *incentives* in the intertemporal decision making processes of the private sector and public sector. This was stressed forcefully by Mäler (1974) and in Solow's reaction (1973) of the Club of Rome report (Meadows et al., 1972). The argument of the Club of Rome is basically that "population growth, production of pollution and consumption of energy - if allowed to continue at rates even quite a bit smaller than present rates - will cause a 'crunch' where society essentially collapses." (Brock, 1977, p. 451). This viewpoint is sometimes referred to as 'neo-Malthusian', due to the analogy with Malthus' perspective that absolute physical

---

[5]In the latter case, no one owns the resource and access is open to all. A common property resource, on the other hand, is owned by some defined group of people, for instance a community or a nation. It is possible, as Pearce & Turner (1990, p.250) stress that "within this group of people there will be open access, that is, each individual member of the group will be permitted to make whatever use they wish of a resource". It is, in my opinion, a matter of taste, whether to view the earth's atmosphere and the ozone layer as a common property resource owned by the world community or as an open access resource, owned by no one. Here, such environmental resources will be referred to as common property resources with open access.

[6]I will not go into the Coase Theorem that states that independent of the property rights, there is a natural tendency to arrive at the social optimum (Coase, 1960).

[7]The tragedy does not mean, as stressed by Dasgupta (1982) that resources are necessarily depleted (see also Chapter 4 below). Rather the equilibrium stock of the resource is (much) lower than the socially optimum one.

limits of land resources will become binding in the foreseeable future (Pearce & Turner, 1990, p.288). The Club of Rome based its conclusions on the famous large scale system analysis world model of Forster *cum suis.* (see Meadows et al., 1972). Solow (1973) heavily criticized this 'Doomsday' model for its lack of ability to capture endogenous price-induced changes in the economic structure.

Solow's trust in the price mechanism as a sufficiently corrective economic feedback can be referred to as (neo-)Ricardian, in analogy with Ricardo's ideas that there are no foreseeable physical limits, but that, instead, the resource stock is non-homogeneous. Price induced technical progress can make 'marginal lands' still productive (Pearce & Turner, op. cit., p. 288). In the case of exhaustible resources, for instance, scarce resources would lead to high prices, giving in turn incentives to shift towards less resource intensive solutions.

In the case of pollution, the role of the price mechanism is equally important. Solow (1973, p. 49) attacks the notion that "air and water and noise pollution are an inescapable accompaniment of economic growth ..". Instead, he argues that excessive pollution happens because of the fact that scarce resources (air, water, etc.) go unpriced, as it is owned by all. However, this flaw in the price system "can be corrected, either by the simple expedient of regulating the discharge of wastes to the environment by direct control or by the slightly more complicated device of charging special prices - user taxes - to those who dispose of wastes in air and water. These effluent charges do three things: they make pollution-intensive goods expensive, and so reduce the consumption of them; they make pollution-intensive methods of production costly, and so promote abatement of pollution by producers; they generate revenue that can, if desired, be used for the further purification of air or water or for other environmental improvements" (Solow, op. cit. p. 50).

As stressed above, for the economic policy discussion, the induced changes are crucial. Intertemporal optimisation models (optimal control models) are typically well suited for decision making processes of the economic agents. An additional advantage of optimal control models is that these enable the modelling of the concept of sustainability in a more or less satisfactory way[8]. Therefore, we concentrate in the literature review and in the subsequent analysis on optimal control models of a whole range of different environmental resources with special focus on Greenhouse models. Note that a drawback of optimal control models is that they are necessarily

---

[8]Also, as said before, these models can straightforwardly be extended to a multi-country framework, where the intertemporal decision making processes are described as multi-actor optimal control models, known as differential games.

very simplified with rather few variables, compared to other model types[9],[10].

The chapter is organised as follows. In Section 2, the Greenhouse Effect is analysed. The literature on environmental resource economics using optimal control models in surveyed in Section 3. Finally, Section 4 looks into the issue of cooperation at the international level with special focus on the incentives leading to a 'tragedy of the commons'. Possible ways to ensure cooperation will also be discussed.

2.2 The Greenhouse Effect

In this section, the background, emission trends and impacts of the (enhanced) Greenhouse Effect are reviewed[11]. The purpose of this exercise is to gain a better understanding of possible economic policy options towards reducing Greenhouse gas emissions. Next, the focus is placed on the sectors in the economy and the types of human activities where major emission reductions may fall. Together these lay the groundwork for the policy discussion in later chapters of how to weigh the costs and benefits of reductions of Greenhouse gas emissions.

2.2.1 Science

Life on Earth is possible because of the *Natural Greenhouse Effect*. This effect results from the capacity of certain long-lived atmospheric trace gases to trap part of the radiant heat which the earth emits after receiving solar energy from the sun[12].

---

[9]See Pezzey, 1989, p. 2 for some critical notes on this issue.

[10]Other modelling types, used in this field include, besides large scale systems analysis models also (non-) linear programming models, econometric models and general equilibrium models. The drawback of linear programming models is that they are not able to capture price-endogeneity appropriately (cf. the RAINS model, Alcamo et al. (1988)). An often-quoted large scale multi-sector econometric model of the U.S. economy is by Jorgenson & Wilcoxen (1990). The novel element is that an attempt is made to model endogenous technical change. An example of a multi-country, multi-sector, dynamic applied general equilibrium model is the GREEN model of the OECD. These and other large scale models are discussed in OECD (1992).

[11]The information used in this section is largely taken from the reports of the Intergovernmental Panel on Climate Change (IPCC, 1990), the report of the Study Committee of the 11th German Bundestag (Enquete Kommission, 1989), the study by the Environmental Protection Agency of the United States (EPA, 1990), the joint report of the Organisation for Economic Co-operation and Development and the International Energy Agency (OECD, 1991) as well as the Greenpeace Report on Global Warming (Leggett, 1990).

[12]The main constituents of the Earth's atmosphere are nitrogen, oxygen, argon and carbon dioxide, accounting for 78.08%, 20.95%, 0.94% and 0.0035% respectively, of the total volume of

Because this phenomenon is somewhat similar to the capacity of greenhouse glass enclosures to trap heat, it is commonly referred to as "greenhouse effect" (Arrhenius & Waltz, 1990). The heat reflection resulting solely from natural greenhouse gas emissions (as in pre-industrial times) raises the Earth's temperature by about 30 °C to an average of 15 °C. For a simplified explanation of the greenhouse mechanism, see Figure 2.1.

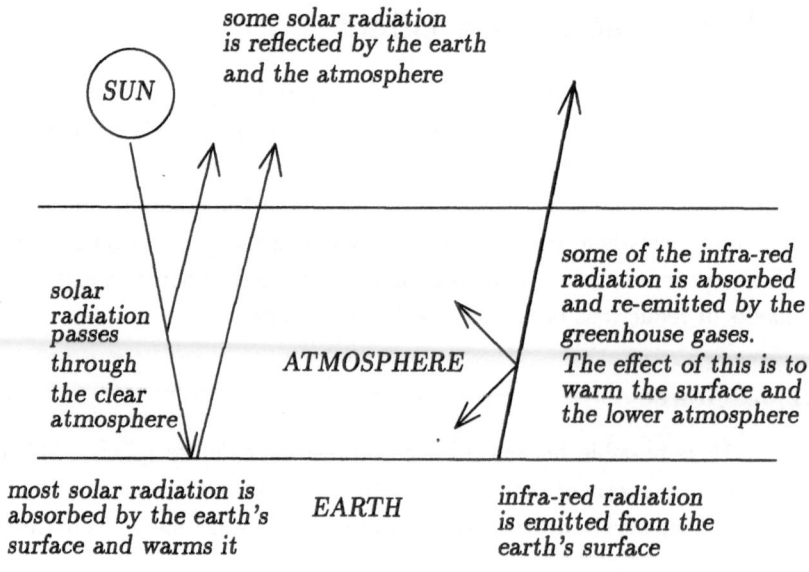

Figure 2.1: *Simplified diagram illustrating the Greenhouse effect; based on IPCC WG I, p. 5; 1990*

This natural greenhouse effect has been *enhanced* since the industrial revolution due to significant anthropogenic (manmade) emissions of especially carbon dioxide (CO_2), methane (CH_4), nitrous oxide (N_2O) and chlorofluorocarbons (CFCs). This *enhanced greenhouse effect* is often loosely called 'The Greenhouse Effect' (GHE). The GHE results on average in an additional warming of the Earth's surface. Water vapour will, in turn, increase in response to global warming and further enhance it.

the Earth's atmosphere. The concentration of these gases is spread out equally. At the same time, the concentration of water vapour varies widely. The latter, together with carbon dioxide are the main natural greenhouse gases.

The consequences of this greenhouse gas build-up are surrounded by uncertainty. This uncertainty is "not about whether the greenhouse effect is real or whether increased greenhouse gas concentration will raise global temperatures. Rather, the uncertainties concern the ultimate magnitude and timing of warming and the implications of that warming for the Earth's climate system, the environment, and economics" (EPA, 1990; p.8). For example, if nothing else changed in the Earth's climate system, it has been estimated that a doubling of CO_2 would raise average global temperature by 1.2–1.3 °C. However, induced increases in atmospheric levels of water vapour and induced changes in the vertical temperature profile and snow and ice covers would eventually lead to an average global temperature rise of 2–4 °C. Additional impacts on clouds and other feedbacks could enhance this warming to roughly 2.5–5.5 °C or diminish it somewhat, perhaps to 1.5 °C (EPA, 1990; p.8).

Although the ultimate implications of global warming are unknown, there is knowledge on past CO_2 and CH_4 concentrations, past temperature fluctuations and their climatic consequences[13]. For instance, total global warming since the peak of the last ice age, some 18,000 years ago, was about 5-8 °C with increases in CO_2 and CH_4 of 50% and 75% respectively (IPCC-WG I, 1990; p.6 and EPA, 1990; p.8). These changes resulted in huge landscape transformations in the Northern hemisphere.

One of the problems with calculating the ultimate mean global temperature rise is the time lag between 'equilibrium warming commitment' and 'realised warming'. The former is defined as "the temperature increase that would occur in equilibrium if the atmospheric composition was fixed in that year" (EPA, 1990; p. 19). Realised warming may considerably lag behind the equilibrium level due to the large heat capacity of the oceans (EPA, 1990; p. 19).

Besides the GHE, global climate change is also potentially affected by other human activities, esp. via desertification, deforestation and changes of stratospheric ozone (IPCC-WG I, 1990; p.6).

2.2.2   Emissions

In order to calculate the relative importance of various greenhouse gases, one has to look at the concentration of the greenhouse gases, their relative effectiveness as a greenhouse gas, their annual growth rate and their atmospheric lifetimes. These

---

[13]The causal link between temperature fluctuations and emissions of carbon dioxide and methane is not beyond discussion.

factors together determine the share of the individual gases in the total effect, as summarised in Table 2.1 and 2.2[14,15].

Summary of Key Greenhouse Gases Affected by Human Activities					
	Carbon Dioxide	Methane	CFC-11	CFC-12	Nitrous Oxide
Units (parts per)	10^6	10^6	10^{12}	10^{12}	10^9
Pre-industrial (1750-1800)	280	0.80	0	0	288
Present (1990)	353	1.72	280	484	310
Current rate of change per year	0.5%	0.9%	4%	4%	0.25%
Atmospheric lifetime (years)	50-200	10	65	130	150

Table 2.1: *Summary of key GHGs affected by human activities;*
(Source:IPCC-WG I)

Relative Contribution of Key GHGs to Greenhouse Effect				
Carbon Dioxide	Methane	CFC 11 and 12	other CFC's	Nitrous Oxide
55%	15%	17%	7%	6%

Table 2.2: *Relative Contribution of Key Gases to Greenhouse Effect;*
(Source: IPCC-WG I)

Note that the atmospheric lifetime of CO_2 as given in Table 2.1 is relatively uncertain, as absorption by the oceans and biosphere is not fully understood yet. Following best guesses of the IPCC, it is assumed that the atmospheric lifetime of

---

[14]Note that only the two most important CFCs are taken into consideration (CFC-11 and CFC-12).

[15]The relative contribution of the different anthropogenic Greenhouse gases is the issue of a lively debate. For instance, the corresponding numbers of Table 2.2 in the often cited EPA-report are: 49% for CO_2, 14% for CFC-11 and CFC-12, 5% for N_2O, 19% for CH_4 and 13% for others (EPA, 1990; p.6). The Greenpeace Report stresses the underestimated importance of carbon monoxide (CO). Its contribution could, according to sources quoted in the report be between 20 and 40% (Leggett, p. 263).

CO_2 is between 50–200 years. Furthermore, note that with current growth trends, carbon dioxide will continue to be the most important anthropogenic greenhouse gas. This will be probably even more pronounced in reality with the rapid phase-out of CFCs, as agreed in the Montreal Protocol of 1987 and the subsequent Amendments in London (1990) and in Copenhagen (1992). Nordhaus (1990) claims that most studies do not fully account for the fact that the various greenhouse gases have different lifetimes and are transformed over time into other substances. He introduces a new index of the importance of different gases that measures the warming impact of a greenhouse gas by its 'total' discounted contribution to warming over the indefinite future (Nordhaus, 1990; p.2). Depending on the discount rate chosen, Nordhaus comes up with a percentage contributions of CO_2 for the period 1985-2100 of 76.13% up to 94.67%. As shown in Table 2.2, the corresponding figure for the period 1980-1990 taken from the IPCC report is 55%. It is difficult to assess to what extent Nordhaus is right in his claims, but even from the IPCC report it is clear that CO_2 is by far the biggest single contributor to the GHE.

For the economic policy discussion, it is not sufficient to know which gases are the most relevant for the Greenhouse Effect. One also needs to know which types of gas-emitting activities are easiest to curb. Attention is therefore turned to a subdivision of the relevant Greenhouse gases (CO_2, CH_4, N_2O and CFCs) by sector of the economy and by type of human activity. In a later section, this information will prove crucial for the discussion of how to slow down the Greenhouse Effect.

Carbon Dioxide (CO_2)

Carbon dioxide is accumulated in the atmosphere mainly through the combustion of fossil fuel (energy). A significant portion of CO_2 is emitted as a result of deforestation and other transfers of carbon between sinks or reservoirs. To a lesser extent, the CO_2 accumulation results from the use of cement in the construction industry. Out of anthropogenic emissions, fossil energy takes 74%, deforestation 25% and cement 1% (IPCC-WG III, 1990; p.5).

Fossil Fuels are especially used for energy generation both in power plants, but also in industry, transport and housing. Note that, even though fossil fuels are often considered together, the carbon content and hence the global warming potential of the various fossil fuels differ considerably. For instance, normalising the warming potential for hard coal to one, the potentials for lignite (soft or brown coal), oil and gas are 1.21, 0.82 and 0.57 respectively (OECD, 1991; p.16; Enquete Kommission,

1989, p.49).

As far as the international division of CO_2 emissions from fuel combustion is concerned, the USA is responsible for about 25%, the rest of the OECD for 23%, Eastern Europe and the Soviet Union for 26%, China for 10% and the developing countries for 16% (IPCC-WG II, 1990; p.8). It is likely that the emission from the later three groups will rise rapidly in the future, compared to emissions in the 'West'.

Transfers of carbon between sinks or reservoirs, from oceans, forests and other growing vegetation, account for a considerable portion of the CO_2 emissions, as mentioned above. The importance of maintaining these reservoirs is beyond doubt. Bolin et al. (1986) estimate that " the airborne fraction of the CO_2 emissions is probably 45 per cent ±10 per cent ... It is also important to note that the more slowly the emissions increase, the smaller the fraction that remains in the atmosphere" (OECD, 1991; p.16). The most important threat to the maintenance of these reservoirs is *deforestation*, which eliminates the possibility of CO_2 being absorbed and stored through photosynthesis (OECD, 1991; p.16).

Methane (CH_4)

As noted in Table 2.1, methane has a much higher potential for absorbing radiation than CO_2. However, due to its considerably shorter lifetime and its much lower concentrations, it does not seem to be as important a greenhouse gas as sometimes claimed. The main anthropogenic sources of CH_4 are enteric fermentation in cattle (20%), releases from rice paddies (30%), biomass burning (14%) , leakage from natural gas (6%) , coal and oil cycles (12%), animal wastes (10%) and solid waste (landfills) (8%) (IPCC-WG III, 1990; p.5). Hence, it seems that the agricultural and energy sector are the main contributors to methane emissions. Besides these manmade sources, CH_4 is also emitted due to releases from wetlands, lakes, tundras and the oceans.

Nitrous Oxides (N_2O)

Emissions of nitrous oxides are for more than 50% due to natural processes, esp. microbial processes in oceans, estuaries and natural soils. The main anthropogenic sources are fossil fuel combustion (57%) , biomass burning (10%), fertilised soils (11%) and cultivated natural soils (22%)[16].

---

[16]The calculations are the point estimates of the numbers given in OECD (1991; p. 18).

Chlorofluorocarbons (CFCs)

CFCs are exclusively manmade. Their (industrial) production started in the 1930s , with a rapid increase from 1950 on till about 1970. Their main use is for solvents, refrigerator fluids, spray can propellants and foaming agents. CFCs are held responsible for ozone depletion, but they are also a major greenhouse gas, due to their enormous warming potential (one molecule of CFCs has about the same climatic effect as 15,000 CO_2 molecules (Enquete Kommission, 1989, p.39). Notwithstanding their rise (5% per year up till recently), the main industrial nations are now rapidly phasing out CFC production, as mentioned above. However, there may be continued rise in the production of some CFCs that are not part of the Montreal Agreement and subsequent Amendments.

Besides the distinction between the types of greenhouse gases it is also worthwhile to look at the emissions of greenhouse gases *by economic sector* or *by type of human activity*. By far the largest anthropogenic source of global warming is *energy production and use*. The "consumption of energy from fossil fuels ... for industrial, commercial, residential, transportation and other purposes results in large emissions of CO_2 accompanied by much smaller emissions of CH_4 from coal mining and the venting of natural gas" (IPCC−*WG* III, 1990; p.1).

The second largest sector is *industry*, especially due to the production and use of CFCs, as well as of other goods etc. The two main other anthropogenic sources of radiative forcing are forestry and agriculture. The first one is due to deforestation, biomass burning and other sources of land-use practices responsible for the release of CO_2, N_2O and CH_4. In the *agricultural sector*, rice cultivation and livestock systems lead to methane releases whereas fertiliser use leads to the accumulation of nitrous oxides. Of the other sectors, CO_2 from cement manufacturing and methane from landfills are the most important emissions. According to the Greenpeace Report (Leggett, 1990; p.169), the estimates for the relative contributions from the sectors energy, industry (CFC), industry (others), forestry and agriculture are: 57%, 17%, 3%, 9% and 13% respectively. In a subsequent section, attention will be on the possibilities of each of the sectors to cut down on its use of GHGs. First, however, possible future trends in the use of GHGs will be discussed.

2.2.3 Future Trends

As shown in Table 2.2 in the last section, the growth rates of most of the greenhouse gases are considerable. Already, at the moment (1990 data), CO_2 levels are more than 25% higher than in pre-industrial times. Of course it is very difficult to predict future greenhouse gas emissions, due to their interrelationship with rates of economic growth, population growth, policy measures taken to curb the greenhouse effect as well as on feedbacks of the greenhouse effect on the economy. Given the difficulty to predict these, most studies look at particular scenarios to evaluate possible future emission trends (e.g. the IPCC and the EPA studies[17]). For instance, the 'Business as Usual' scenario of the IPCC - Scenario A - assumes a continuation of coal intensive energy supply with only modest efficiency increases. Deforestation continues and CFC substitution strategies are implemented according to the Montreal Protocol albeit with only partial participation. This implies a continuation of the growth rates for CO_2 as given in Figure 2.2 and a very moderate slow-down of other emissions, leading to a doubling of equivalent CO_2 concentrations compared to pre-industrial times in about the year 2025. (IPCC-Overview, p.1). Similarly, the EPA study predicts for corresponding calculations a doubling in the year 2030-2040 (EPA, 1990; p.1).

The other three scenarios (B,C & D) of the IPCC report are characterised by stepwise increasing levels of controls[18].

In all scenarios of both EPA and IPCC, it is assumed that the developing countries 'play their part'. If developing countries do not participate in a global greenhouse agreement, equilibrium warming is about 40% higher in 2050 for the 'Stabilising Policy' cases, with risk of substantial global warming now being unavoidable.

Of course, it is questionable to what extent the scenarios of the EPA and the IPCC studies are realistic, especially because it is assumed that no endogenous

---

[17]The IPCC and EPA studies explore the climatic implications of four resp. six different scenarios of future patterns of economic development, population growth and technological change. Each of the scenarios are claimed to be internally consistent.

[18]The most radical scenario (D) assumes a shift to renewable and nuclear energy in the first half of the next century. This leads to a reduction of carbon dioxide emissions of 50% of 1985 levels by the middle of the next century (IPCC-WG I, 1990; p.26). Besides, CFCs are phased out rapidly and agricultural emissions are limited. This scenario shows that stringent controls in industrialised countries combined with moderate growth of emissions in developing countries could stabilise atmospheric concentrations at twice the pre-industrial level towards the end of the next century. Only even more radical steps (an immediate 60% reduction of net emissions) would achieve stabilization of concentration at today's levels (IPCC-WG I, 1990; p.26).

The most radical EPA scenario assumes high carbon emission fees ($19/barrel on oil), rapid reforestation and drastic rates of technical efficiency improvement. This leads to a global decrease in CO_2 equivalent emissions of 60% in 2025. Only in this last, aggressive, policy case will the greenhouse gas build-up be reversed early in the 21st century. Equivalent CO_2 concentrations will be back to current (1990) levels in about 2100 (EPA, 1990; p.20).

behavioural changes occur and that no economic feedbacks of global warming take place. This implies that the forecasts of the 'Business as Usual'/'No Response' scenarios may be unrealistic. Basically, the same (neo-)Ricardian criticism applies to these models as to the 'Doomsday model' that the Club of Rome used.

2.2.4 Impacts

Notwithstanding vast uncertainties, a thorough description of possible impacts is crucial for economic policy discussions. The reason is that the monetary valuation of such impacts corresponds to the costs of not taking policy measures. Put differently, in a cost-benefit framework of the Greenhouse Effect, avoidance of impacts forms the benefits of measures to curb GHG emissions. In this section, the impacts will first be described in physical terms. These impacts can be assessed, once the concentration of GHGs (the 'dose') as well as the dose-effect relation are known. At the end of this section, the possible monetisation of the impacts will be discussed.

The predicted effects of greenhouse gas emissions include not only a rise in the *global mean temperature*, a rise in *global sea levels*[19] and a shift in *global patterns of wind and ocean currents*, but particularly also the increased *risk of drought* and an increase in the possibility of *extreme weather events*, which are non-linearly correlated with mean temperatures (Arrhenius & Waltz, 1990; p.7). The latter can lead to 'flip-flops' of the climatic system which are extremely difficult to predict. Therefore, "we may be 'lulled into complacency' by model simulations suggesting a gradual warming over the next century" (Arrhenius & Waltz, 1990; p.7).

As far as signs of global change up to now are concerned, it is supposed that "Global mean surface air temperature has increased by 0.3 °C to 0.6 °C over the last 100 years, with the five global mean warmest years being in the 1980s. Over the same period global sea level has increased by 10-20 cm. These increases have not been smooth with time, nor uniform over the globe ... There is no firm evidence that climate has become more variable over the last few decades" (IPCC-WG I, 1990; p.3). These estimates of mean temperature rise are consistent with the data given in the studies of the EPA and the Enquete Kommission. However, the proof of

---

[19]Estimates of sea level rise vary widely. The IPCC Business-as-Usual scenario is alleged to give a rise of 65 cm by the end of the next century (IPCC, WG-I, 1990; p.2). However, in the 'Potential Impact Study' of the IPCC, a sea level rise of 100 cm is assumed (IPCC, WG II, 1990; p.8). Finally, the Enquete Kommission predicts a rise of 150 cm if emission patterns do not change, although it does not exclude the possibility of a 500 cm increase (Enquete Kommission, 1989; p.41-2).

whether there are already signs of global changes are widely disputed as the observed changes need not necessarily be related to greenhouse gas emissions. Proponents of global warming claim that the climatic change has not been so pronounced up till now due to the fact that realised temperature changes may lag considerably behind the equilibrium level of warming, as explained above.

The climatic changes will through biological and physical effects ultimately have socio-economic consequences (IPCC-Overview, 1990; p.4). The IPCC study gives potential impacts of an effective doubling of CO_2 in the atmosphere between now and 2025 to 2050, a consequent increase of global mean temperature in the range of 1.5 °C to 4-5 °C and a sea-level rise of about 30–50 cm by 2050 and about 100 cm by 2100. The consequences of such changes are extensively dealt with in the IPCC-Overview for the categories agriculture and forestry (1), natural terrestrial ecosystems (2), hydrology and water resources (3), human settlements, energy, transport and industrial sectors, human wealth and air quality (4), oceans and coastal zones (5) and seasonal snow cover, ice and permafrost (6) (IPCC-WG III, 1990).

For instance, for the sector agriculture and forestry, the study indicates that Climate Change will have important effects on agriculture and livestock. However, it has not yet been conclusively determined whether, on average, the global agricultural potential will increase or decrease. It seems that in regions with high vulnerability (e.g. Sahel, China) there may be severe negative effects. On the other hand, there is scope for productivity increase due to prolonged growing seasons in countries on high and mid latitudes. Hence, countries like Canada and Russia may be net benefitters from climate change.

For oceans and costal zones, it is assumed that an increase in the sea level of 100 cm would render some islands and coastal zones uninhabitable, displacing tens of millions of people. Changes in the heat balance of oceans may have various severe effects, such as a decrease in bio-diversity and decrease in fishery yields.

The cost of *adjustment investment*, needed in order to cope with sea level rise (dams, etc.) may be severe and will divert investment away from directly productive purposes.

Summarising, reasonable rises in the global temperature could lead to moderate but also to dramatic and irreversible effects. Especially uncertain is the extent to which ecosystems and people can adjust to predicted levels of global warming. Besides, the likelihood that 'flip-flops' of the climatic system occur is not known with any degree of certainty. The impacts may especially be pronounced in developing countries. This complex of uncertainties makes a *monetary evaluation* extremely tricky. Some economic categories are relatively easy to measure and inherently monetizable, because there exists a clear market value (e.g. crop losses). Another example where monetisation might be possible, is coastal sea management (Nordhaus, 1990).

Other possible impacts are, however, "difficult to quantify and can only be weighed using some mix of quantitative and qualitative analysis. Still other categories such as preservation of coastal wetlands, old growth forests and biological diversity, with their attendant option and existence values, are difficult if not impossible to quantify. All of which argues for more research on benefits and a plea to avoid the simple conclusion that if we cannot measure the benefits with precision, they effectively have zero or low value to society" (Morgenstern, 1991, p.142-143).

One bold attempt to assess the overall impact of global change for the U.S. economy is by Nordhaus (1990) who claims that most sectors of the industrialised economies have no direct interaction with climate and impacts of GHE on these sectors are likely to be very small. He assumes that only those sectors of the economy that have a significant interaction with unmanaged ecosystems – that is, are heavily dependent upon naturally occurring rainfall, run-off, or temperatures – may be significantly affected by climate change. Agriculture, forestry, and coastal activities fall in this category (Nordhaus, 1990). Coastal activities are likely to be influenced in a negative way, whereas agriculture and forestry may be affected both negatively (drought, etc.) and positively (esp. due to the fertilising effect of higher CO_2 concentrations).

This assessment is dramatically different from the one of the IPCC[20]. The reason is that Nordhaus concentrates on the USA and does not take into account economic sectors for which the impacts are difficult to quantify, nor the potential dramatic breakdown of ecosystems. His claim is that the USA and other industrialised countries should be able to cope quite flexibly with the climatic changes, given their long time frame (mountain skiing is replaced by water skiing, agriculture in the US moves towards the North, etc.).

In conclusion, the consequences of global climate change are highly uncertain, with most probably a net overall economic loss, though some countries may be better off. The negative effects may be much more pronounced in developing countries

---

[20]Morgenstern (1991) notes that correcting for a computational error of protecting open-coasts and taking some other issues into account leads to an approximate doubling of Nordhaus's annual costs estimate.

and, more generally, in countries with large climate-sensitive sectors. Besides, the effects may depend very much on the speed of the global change. Really catastrophic events cannot be excluded, even with a moderate Greenhouse gas build-up.

2.2.5   Policy Options for Limiting the Impacts

There exist many policy options for limiting the impacts of Greenhouse gas emissions and it is worth noting that "no single activity is the dominant source of greenhouse gases; therefore, no single measure can stabilize global climate. Many individual components, each having a modest impact on greenhouse gas emissions, can have a dramatic impact on the rate of climate change, when combined" (EPA, 1990; p.19).

At the same time, however, it is worth concentrating on some of the most powerful options. As shown in Table 2.2, CO_2 and CFCs are the two biggest contributers to the GHG build-up. As for CFCs, it seems that the Montreal Protocol and the London and Copenhagen Amendments have paved the way for a almost complete phase-out of CFCs within the foreseeable future. Fear for ozone depletion and not global warming were the reason for the consensus. In fact, cost-benefit analysis of ozone depletion alone justifies an almost complete phase-out of CFCs (Barrett, OECD, Ch. 3, p.134, 1991). Hence, CFC policy options for the Greenhouse effect will not be considered.

Therefore CO_2 will hence be the focus of attention[21]. As noted above, the main activities leading to CO_2 emissions are energy production and use (74%) and deforestation (25%). Hence the policy options of interest are policies concerning energy production and use including technological progress and policies concerning the slow down of deforestation and the stimulation of reforestation. These two policies will be discussed below.

---

[21]Methane (CH_4) is the third largest contributer to the Greenhouse Effect. However, it is not easy to lower these emissions sizably. For example, stabilising GHG emissions at current levels in a cost-effective way implies cutting CO_2 emissions with at least 50%, while methane is only supposed to decrease with 10%. The reason is that there are no easy options for the main anthropogenic sources of methane emissions (releases from rice paddies, 30%; fermentation in cattle, 20%), as explained above.

As for N_2O, due to its relatively small contribution to the Greenhouse effect, this source will not be considered.

Policies concerning Energy Production and Use

As energy production and use are by far the largest single source of greenhouse gas emissions, "the current energy supply system has to be subjected to a fundamental review in order to identify ways and means of minimising or eliminating adverse effects on both man and nature" (Enquete Kommission, 1989; p.44). Note that apart from the energy sector itself, major amounts of energy are used in the sectors industry, transport and housing. Policy options to decrease CO_2 emissions in production and use of energy include:

- improvement of energy efficiency in all sectors, for instance use of energy-saving technology, super-conductive materials, advanced heat-pumps, improved energy storage, etc[22];

- other forms of energy conservation, for instance improved insulation;

- other forms of lowering energy demand, for instance through energy pricing that includes the external costs[23], and through information and environmentally conscious behaviour (Enquete Kommission, 1989; p.50), etc;

- fossil fuel shift. Given the lower carbon content of gas and oil compared to soft and hard coal, a shift away from coal will decrease emissions considerably;

- shift towards non-fossil fuels, for instance renewable energy sources (biomass, solar and wind energy) as well as nuclear energy;

- stimulation of technological innovation and control of the combustion process and energy transport.

Each of these options can contribute to moderate reductions of fossil fuel use. However, taken together, these measures can result in dramatic decreases in the GHG build-up. Ideally, these measures should be taken in such a way that the marginal costs are equated over different options and over the various regions in the world. This will be analysed in the next section. First, however, other policy options will be discussed.

---

[22]A detailed list of options is given in the OECD/IEA study (OECD, 1991).
[23]This was stressed by the IPCC report (Overview, 1990; p.11), as well as by the Enquete Kommission (1989, p.50).

Policies concerning Forestation

Through natural photosynthesis, carbon dioxide can be stored in terrestrial and aquatic vegetation. Deforestation leads to a decrease in the total storage possibilities. Likewise, new CO_2 reservoirs can be created via reforestation[24]. Note, however, that only in their growth-phase do trees absorb carbon dioxide. Fully developed trees, on the other hand, form only a storage of CO_2, but they do not capture net quantities of CO_2. Hence, lasting increases in the absorbtion rate of CO_2 by trees is only possible with ever increasing reforestation.

Deforestation practices often involve the burning of massive amounts of wood, releasing major amounts of CO_2 (and CH_4). Given the fact that currently deforestation is kept responsible for 25% of CO_2 emissions, reversing this deforestation could be quite an important option. Note however, that large areas are needed for reforestation to have a substantial impact. Nordhaus (1991) concludes for instance that the quantitative importance of forest harvesting at these modest efficiency costs is quite small. Besides, reforestation attempts must compete against alternative land-use demands within the biosphere (Arrhenius & Waltz, 1990; p.14). On the other hand, most deforestation is economically inefficient and replacing existing forestry management by a sustainable one would already lead to major decreases in carbon releases.

Other possible future policy options might include such things as abatement of CO_2 via scrubbing and geological and marine disposal of CO_2. Arrhenius & Waltz (1990; p.13) and the Enquete Kommission (1989, p.49) are quite sceptical about their economical feasibility, though the OECD/IEA study (OECD, 1991; p.94-5) is less pessimistic. Both options would, however, only become potentially interesting in the very long run. Given the uncertainty of their feasibility, these 'abatement' options will not be considered in the rest of this dissertation.

A final policy option concerns population growth. Given the obvious positive correlation between population size and energy use, a global population policy seems an obvious, though controversial, choice (IPCC-Overview, 1990; p.11). This is especially true for regions that are already ecologically vulnerable, due to desertification, deforestation and non-sustainable agriculture, often as a result of population growth pressures. Although this may be an extremely important issue, it will not be elaborated further.

---

[24]For instance, the American Forestry Association launched a plan in 1988, called Global Re-Leaf, aiming to encourage tree planting in US-communities.

Up till now, only options to limit the concentration of trace gases in the atmosphere have been described (preventive measures). The impact of global warming can, however, also be reduced by measures for adjusting to potential climate change (adaptive measures). Two broad categories of adaptation options are *coastal zone management* and *resource use and management* (IPCC-Overview, p.22-7).

Coastal zone management, especially in vulnerable areas, such as the river deltas of Bangladesh, Egypt and Vietnam, could range from retreat and accommodation (flood shelters) to protection (sea walls, etc.).

Resource use and management include better water management practices, more sustainable agricultural practices as well as the use of more resilient resources.

Given the complementarity of the policy options outlined above, it is "necessary to consider limitation (i.e. preventive, ed.) and adaptation strategies as part of an integrated package in which policies adopted in the two areas complement each other so as to minimise costs" (IPCC-WG III; p.22). Adaptive measures have two advantages, as Barrett (1991 OECD Ch.3, p.55) notes: adaptation does not require international cooperation and besides discounting will favour adaptation[25].

Ideally, one would like to have estimates of the costs of different policy options per unit of CO_2-equivalent reduced, so as to rank a whole range of different strategies in their cost effectiveness. This has been done by, among others Nordhaus(1990), Manne & Richels (1990), Jorgenson & Wilcoxen (1990) and Barrett(OECD, Ch.3 1991). For a comprehensive comparison of cost estimates, see OECD (1992). Nordhaus (1990) gives a comparison between costs of different forms of CO_2 reduction, reforestation and CFC reductions for the U.S. His estimates "suggest that a significant reduction of greenhouse gas emissions can be obtained at modest cost. It is possible to reduce greenhouse warming by about 15 percent for a cost of less than \$ 10 per ton CO_2 equivalent ..." (Nordhaus, 1990; p.30). Other studies claim that considerable reductions can even be achieved at a profit, as the investment in energy efficiency and sustainable land use are often justifiable in its own right[26]. (NAS, 1991).

---

[25]The adaptation options will not be given much attention in the rest of this dissertation, as they are rather difficult to model explicitly in the type of models used.

[26]Arrhenius & Waltz quote a study in which "it was demonstrated that a 10 billion investment in cost-efficient improvements in electricity end-use could offset anticipated demand for 22 GW of new generating capacity. The comparable capital cost of having to install additional capacity of an equivalent amount would be on the order of 40 billion" (Arrhenius & Waltz; 1990; p.12)

2.2.6   Policy Instruments

In order to achieve the best results at minimal costs, the choice of policy instruments
is crucial. In the environmental economics literature, two types of instruments are
generally distinguished[27]: the 'command-and-control' approach and the 'economic
incentive' (or 'market-based') approach (cf. Bohm & Russell, 1985). Examples of
the latter are carbon taxes, subsidies on carbon reductions and tradeable permits.
The optimal tax (subsidy) on carbon (carbon-reductions) or, in the case of permits,
the price of emission certificates, will equal the corresponding per unit cost of
'pollution'.

As noted before, the impacts of the Greenhouse Effect cannot be easily mone-
tised. It may therefore be that the optimal level of emission reduction cannot be
calculated. However, even then is it optimal from an efficiency point of view to
use economic incentive instruments. Cost-effectiveness can be obtained by equating (through taxes, permits, etc.) the marginal 'abatement' costs for each type of
Greenhouse gas and for each type of measure (preventive and adaptive) and for
every region in the world[28]. In the case of fossil energy, for instance, it is therefore
important to look at the carbon content of the different fuels. Only when economic
efficiency is achieved, are total 'abatement' costs minimised for a given target level
of emission reduction. "A 'command-and-control' approach or fiscal instruments
which do not take account of the carbon content of fuels are likely to increase the
macroeconomic cost significantly compared with approaches that operate through
economic incentives: carbon taxes and emission trading. Both are likely to achieve
emission constraints at least cost" (OECD, 1992).

, A further benefit of market-based measures is 'dynamic efficiency'. By this is
meant the continuing incentive to search for innovative ways to further reduce emissions. As the IPCC report puts it, the advantage is that economic incentive-based
approaches "through their encouragement of flexible selection of abatement measures, tend to encourage innovation and the development of improved technologies
and practices for reducing emissions and therefore frequently offer the possibility
of achieving environmental improvements at lower costs than through regulatory

---

[27]The practice of gentlemen's agreements between government, industry and other interest
groups may be seen as a third approach with a firm stress on negotiated agreements; it can
include both juridical and economic-incentive measures.

[28]Note that in the case of the GHE, there is a one-to-one relation between emission and world
wide GHG build-up. Therefore equal carbon taxes are efficient. For acid rain problems, where
the location of emission is crucial for the impacts, efficiency implies that 'acid taxes' should differ
with the ambient impact it causes (cf. Mäler, 1989).

mechanisms. It is not likely, however, that economic instruments will be applicable to all circumstances" (IPCC-Overview, 1990; p. 12).

Besides the efficiency criteria, equity arguments should be considered, as the distribution of welfare-changes across sectors and regions differs widely with the policy instrument chosen. Ideally, the efficiency and equity issues are dealt with separately, in order to achieve first-best welfare solutions. In this case, a globally cost-effective strategy to curb Greenhouse gas emissions through carbon taxes could be accompanied by an 'equitable' international redistribution of the tax receipts . Likewise, a 'fair' initial allocation of emission certificates can guarantee cost-effective emission reductions. Note that the implicit value flows in the case of a tax redistribution and permit allocation are identical, and hence the equity considerations are exactly the same for both cases.

As seen in the last section on policy options, the distinction between short-run and long-run 'abatement' possibilities is fundamental. "In the short term, the most effective means of reducing emissions is through strategies that rely on pricing and regulation. In the long term, policies to increase research and development of new technologies and to enhance markets through information programs, government purchases, and other means could also make a major contribution" (EPA, 1990; p.33).

The most obvious means to encourage energy efficiency *in the short run* would be to ensure that prices of energy and end-products reflect their full social costs (carbon taxes). This is the more true as low tariffs for electricity, especially for industrial consumers, present a strong disincentive for conservation investment (Arrhenius & Waltz, 1990; p.12).

In the long run, forceful additional instruments would be public education and information as well as subsidising the research of clean and efficient technologies. In the latter case, diffusion of state-of-the-art technology seems to be of the utmost importance.

In Chapter 3, short term as well as long term measures will be dealt with. This is done by introducing both carbon taxes and subsidies on energy saving technologies. The tax receipts from the emission charges are ear-marked for subsidising research of clean and efficient technologies.

The distinction between short and long run is also crucially important as far as technical and economical feasibility is concerned. The reason is that many energy saving techniques and renewable energy projects will only in the long term become feasible. Given the large investment involved, a clear strategic long run government policy is necessary.

Feasibility is, given the global dimension of climate change, also enhanced by *transfers of technology*, especially to developing countries. This is especially important as a considerable Climate Change seems to be unavoidable without the cooperation of developing countries. This issue is dealt with in Chapter 5.

In order to further increase the feasibility of economic instruments, they should, where possible, be implemented globally, so as to avoid distortion of competition. Acceptability within each country is enhanced, if the measures are accompanied by changes in income and corporate tax so as to avoid a too heavy burden on lower income classes (Bleijenberg, 1990).

Before the points mentioned above are worked through in the following Chapters, the existing literature on optimal control models of the environment will be reviewed.

2.3   Environmental Resource Models

2.3.1   Introduction

The aim of this section is to find an appropriate modelling framework for the relevant decision making processes underlying Greenhouse gas emissions using optimal control techniques. Therefore, within the literature on natural resource economics using optimal control models, the focus will be on environmental resources, where the stock of pollution or waste is the central state variable. From the last section it is fair to conclude that Greenhouse policy should concentrate especially on reducing fossil fuel use and stopping deforestation. The former can be obtained by lower energy demand, stimulating energy-related technology and shifts to renewable energy sources. Hence, models highlighting these points especially will be analysed. This helps in:

- defining which elements are lacking in current Greenhouse models;

- defining which elements in other environmental resource models on waste and pollution are helpful for the modelling purposes in later chapters.

Before reviewing environmental resource models, it is worthwhile to note that the literature on exhaustible resources (Hotelling, 1931), renewable resources (Gordon, 1954) and environmental resources using optimal control techniques have de-

veloped quite separately from each other. This is the more surprising given the
analytic similarities between the three, as stressed by Smith (1977) and Dasgupta
(1982) among others. Take, for instance, a very simple optimal control model,
where a social planner is supposed to solve the following optimisation problem[29]:

$$\max_{C} \int_{0}^{\infty} e^{-\rho t} U(C, S) dt \qquad\qquad \rho > 0 \qquad\qquad (2.1)$$

$$\text{s.t.} \qquad C \leq q(R) \qquad\qquad (2.2)$$

$$\dot{S} = kG(S) - \alpha f(R) \qquad\qquad (2.3)$$

In this model, C is consumption and S (resp. R) the stock (resp. flow) of a natural
resource. The functions $U(.)$, $G(.)$, $q(.)$ and $f(.)$ denote social welfare, the growth
of the stock of natural resource, the production in society and the use of natural
resources respectively.

In the case of *exhaustible resources*, k is zero and $\alpha f(R) = R$ is the flow of
extraction of the natural resource (e.g. fossil fuel) used for the production of $q(R)$
unit of output.

With *renewable resources*, k equals unity and $\alpha f(R) = R$. The function $q(.)$
denotes the production of resource-related goods (fried cod's roe, etc.) in a society
just consisting of fishermen who catch R units of fish per unit of time.

In the case of *environmental resources*, the stock S is the 'environmental quality'.
Defining $\bar{S}$ as the environmental quality in the 'virgin state of nature' (the times of
Adam and Eve), it is clear that the current 'stock of pollution', P, equals $\bar{S}$ minus
the current level of environmental quality. Therefore, also $\dot{S} = -\dot{P}$. It is often a
matter of taste whether to model environmental resources using S or P.

More importantly, there are two basically different types of modelling environmental resources. Firstly, R can be considered as an input. In that case, $\alpha f(R)$
equals R, just as in the renewable and exhaustible case[30]. Note that R can either
be thought of as a productive input, such as fossil fuel and fertiliser, or alternatively
as a joint production factor, such as waste. The elegance of modelling pollution
as a joint product is that it allows for exactly the same format as used in renew-
able/exhaustible resources models.

---

[29]Capital and labour are assumed fixed.

[30]In the case of an exhaustible environmental resource, such as the ozone layer, k is zero and
for renewable environmental resources $k \neq 0$. The expression $kG(.)$ is often referred to as the
assimilation of pollution by the environment.

A second way to model environmental resources is by taking $q(R) = \bar{q} = f(R) = \bar{f}$. This way is most frequently used in optimal control models on pollution. Production is here fixed and $\alpha\,\bar{q}$ is the pollution inherent to production. I will refer to these two interpretations as the 'polluting input model' and the 'polluting output model' respectively. In Keeler, Spence & Zeckhauser (1971), the pollutants corresponding to these two types of models are described as 'pollutants as intermediate goods' and 'pollutants as by-products of production'.

Smith (1977) stresses that in the case of renewable resources, there always exists a level k low enough for the optimal trajectory to result in extinction[31,32]. Dasgupta (1982) points out forcefully that "a central common characteristic of such (environmental (ed.)) resources is their regenerative capacity, and therefore, that environmental-pollution problems ought to be thought about in much the same way as ought problems, associated with the control of land, fisheries, aquifers and forests" (Dasgupta, 1982, p.ix). It has been stressed above that this is not always the case (e.g. ozone layer). The presence of irreversibilities in many pollution problems implies some kind of exhaustibility. In the case of the Greenhouse Effect, on the other hand, the atmospheric life-time of, for instance, carbon dioxide is 50–200 years. It is assumed that there are no irreversibilities. This means that the atmosphere (and hence the Greenhouse Effect) can be described as a renewable resource[33].

The similarities pointed out above will appear to be of considerable importance in interpreting the models of environmental resources, to which attention will be directed here.

The literature on environmental resources using optimal control techniques includes topics like industrial waste (with possibly recycling possibilities), air pollution problems (acid rain, ozone depletion, Greenhouse Effect) and wilderness use. Here the focus is on the first two, summarising Keeler, Spence and Zeckhauser (1971), Plourde (1972), Smith (1972, 1977), Forster (1972, 1975, 1977), D'Arge & Kagiku (1973), Mäler (1974), Brock (1977), Dasgupta (1982), Luptáčik & Schubert (1982), Nordhaus (1982;1990), Barbier & Markandya (1989), Cesar (1989, 1990),

---

[31]This is the case if the marginal rate of consumption does not go to infinity (negative) when the level of consumption approaches zero.

[32]Hence, "non-renewable (or: exhaustible, ed.) resources and the problem of exhaustion are just limiting (zero growth) cases of renewable resources, and the problem of species extinction. Just as exhaustion is optimal, extinction can be optimal" (Smith, 1977, p.1).

[33]CFCs also have a finite atmospheric life-time. However, most of the degrading process of CFCs is due to reactions with ozone, through which the layer depletes.

Musu (1990, 1992), Tahvonen & Kuuluvainen (1990), Hoel (1990), Pethig (1990), Gottinger (1990, 1991) and Van der Ploeg & Withagen (1991).

These models are all deterministic in nature and analysed with Pontryagin's maximum principle.

The most elementary models, capturing the intertemporal trade-off between consumption and pollution/waste are very similar to the general natural resource model just described. Nordhaus (1982), in his Greenhouse model[34] takes for instance:

$$\max_C \int_0^\infty e^{-\rho t} U(C) \, dt \qquad \rho > 0 \qquad (2.4)$$

$$\text{s.t.} \qquad C = f(F, P) \qquad (2.5)$$

$$-\dot{P} = \alpha P - \gamma F \qquad (2.6)$$

In this model[35], F stands for fossil energy (both its use and the corresponding emissions) and P is the stock of pollution. As pointed out above, $-\dot{P} = \dot{S}$ where S stands for environmental quality. Note that in this model, the emissions can be seen as an input. In the terminology introduced above, it well be referred to as a 'polluting input model'.

A crucial difference between the two models above is that Nordhaus (1982) assumes that pollution does not enter the utility function directly. In Pezzey's parlance, the stock of environmental resources has no *amenity* value but only a (negative) *productivity* effect (Pezzey, 1989). Hence, the dynamic trade-off between consumption and pollution is only brought about by the negative impact of fossil energy use on future production.

This model of Nordhaus (1982), however incomplete it may seem, captures at least three essential elements:

- the intertemporal trade-off between consumption and pollution;

- the idea of sustainability (via Equation 2.6)[36];

- the idea that a social optimum can be reached via emission taxes.

---

[34]Nordhaus (1982) has a slightly more complicated model, as they assume the production function to be non-continuous, depending on a critical level of Greenhouse gas accumulation. This point will be not be considered here.

[35]f(.) is assumed to be increasing in F, decreasing in P and concave in both arguments

[36]The sustainable level of pollution is then $P = F/\alpha$, as will be elaborated below.

The last can be seen by decentralising Model 2.4 – 2.6 into a model with a representative producer, a representative consumer and a government (tax-collector) who sets taxes, τ. Assume that the producer maximises profit, π, taking into account emission charges by the tax collector: $\pi = pf(F, P) - \tau F$, where p is the commodity price. Assume furthermore that the consumer maximises consumption given his budget constraints $C = \pi + T$, where the government redistributes taxes $T = \tau F$.

In this case, under perfect foresight and full rationality, the social optimum will be reached if τ is set equal to the shadow price of pollution, i.e. the co-state variable of the optimal control model in utility terms.

The major shortcomings of Nordhaus' (1982) one-state variable model are:

- the notion of *sustainability* is not realistically modeled, nor is the concept of *irreversibility* taken into account;

- *abatement* and *other forms of emission reduction* are not possible: the only way to decrease Greenhouse gas emissions is to produce less;

- capital and labour do not enter explicitly in the model and neither *capital accumulation* nor *population growth* are taken into account;

- the possibility of *endogenous technological progress* through investment in human capital is not taken into account;

- the *international* dimension is ignored as well as other aspects that give rise to *strategic behaviour* (interaction between the private sector and the government);

Models of pollution, waste and Greenhouse gas emissions dealing with these four elements will be discussed in the following sections.

Besides these issues, there are inherent weaknesses in the use of optimal control models and of neo-classical analysis in general. I would like to briefly focus on three of them: the concept of rationality, the modelling of uncertainty and the issue of intertemporal equity.

Rationality

It is assumed in the cited papers, that a monolithic social planner maximises welfare in an infinite time horizon setting. The results can be decentralised by replacing the social planner by a market economy with a government, and a representative consumer and producer. Many questions arise:

- Do governments and consumers have an infinite time horizon?

- What happens if preferences of governments and consumers differ?

- What if a government consists of different actors (public choice)?

- Is it fair to assume that economic agents optimise their behaviour or at least act as if they do?

- What about myopic governments and long-term ecological risks?

- Do consumers really deal with long-term issues of environmental degradation in a rational way at all (compare smoking behaviour)[37]?

These and many other questions undermine of course any policy conclusions drawn from optimal control analysis.

These issues, however interesting they are, will not be further worked out, but should be borne in mind nevertheless.

Uncertainty

The models analysed below use a deterministic setting. This means that the decisions are made under the assumption of perfect foresight. However, it is exactly the question of uncertainty, in the case of the Greenhouse Effect for instance, that causes problems of policy-making.

First of all, as noted before, the actual level of GHG-emissions is unknown. Next, the relationship between emissions and GHG-concentration is not clear. Furthermore, the link between increased concentration and the impacts is not known with any degree of certainty. Finally, the monetisation of the impacts is, at the very least, open to wide discussion.

This urges for an explanation of why uncertainty is not dealt with in these problems where uncertainty is at the very core of the problem[38]. The main explanation

---

[37]The possible non-rationality of the public at large in confronting long term environmental damages is, in my opinion a very interesting issue. Quite some work in this area has been done in the psychological literature on this. Most of this is linked to the problem of uncertainty and risk (cf. Tversky, Slovic & Kahneman, 1990)

[38]A model on the Greenhouse Effect taking uncertainty into account is the PAGE-model (Hope, 1991).

is that uncertainty drastically complicates the analysis and that the deterministic setting can at least function as a benchmark case. Different forms of uncertainty can be compared with this case. Siebert (1992) analyses, for instance, how different forms of uncertainty affect the policy results vis-á-vis this benchmark case. His main conclusions are[39] that increased uncertainty in the damage function typically implies a lower level of pollution and that risk aversion strengthens this conclusion. Furthermore, uncertainty in the assimilative capacity and in the impact leads to further decreases in the optimal level of pollution[40,41]. Each of these forms of uncertainty can, in fact, be analysed to an extent using *stochastic* control theory. The problem with solving stochastic control models is that explicit analytical solutions can only be obtained in very restricted settings.

Therefore, in this dissertation, the issue of uncertainty will not be dealt with explicitly. Instead, in Chapter 3, as an example, variations in the specification of the assimilative capacity will be analysed in a deterministic setting. This gives an indication of the importance of uncertainty about the assimilation function for policy purposes. The conclusion of this exercise is that slight variations can lead to enormous changes in the steady state values of pollution and consumption. Therefore it would be crucially importance for future research in this area to model uncertainty explicitly.

Intertemporal Equity

In Nordhaus' (1982) model discussed above, the social welfare function is maximised over time. Future and current welfare are compared by discounting at a constant annual rate $(1 + \rho)$. Two questions arise:

- Is it ethical to discount over generations?

- If so, what is the appropriate discount rate?

---

[39] This summary is based on Pezzey (1989).

[40] With regards to the risk of irreversibility, an option-value could be introduced to explicitly handle the fact of retaining the option of future use of an environmental resource (see also Pethig (1990) and the references therein).

[41] An interesting paper in which uncertainty is explicitly modeled in the context of learning in time about environmental pollution (in his case the Greenhouse Effect) is by Kolstad (1992). He assumes that there is both irreversibility in the assimilative capacity of nature as well as in the abatement investment. The latter urges for lower abatement than without the irreversibility and the former for higher. Therefore, it is not clear, in general, whether introduction of irreversibility would induce policy makers to be more prudent with nature or less.

There is no definite answer to the first question. The Rawlsian approach to this question is that "a fair allocation is defined as one resulting from a decision rule agreed upon by all parties affected by the decision, from a position in which they cannot be sure which way any decision will affect them. This is the 'veil of ig-. norance' procedure" (Hanley, 1992). Rawls (1974) did not intend to apply it to intergenerational allocations, though many have used it this way (e.g. Solow, 1974; Becker, 1982). This leads, with risk-averse individuals, to a maximin rule, in which the optimal discount rate is zero, so that there is no discounting.

The Paretian approach (in the Kaldor-Hicks definition), on the other hand, uses positive discount rates for comparison between generations. The reasons for discounting (pure time preference, productivity of capital, risk, etc.) are well known and will not be analysed here. The main question is one of determining the appropriate level of the social discount rate. Generally, it is assumed that this rate is lower than the private discount rate (see Pearce & Turner, 1990), but there is no agreement about the optimal level itself.

In this dissertation, the Paretian perspective (in the Kaldor-Hicks definition) is taken throughout. However, as the optimal level of ρ is not known in practice, the consequences of changes of the social discount rate are analysed by simulation in the models developed below ($\rho = 0.03$, $\rho = 0.01$ and $\rho \downarrow 0$). The conclusion is that relatively small changes in the level of ρ lead to considerable variation in the key policy variables.

The simplifying assumptions taken here with respect to rationality, uncertainty and intergenerational equity should be borne in mind in what follows. The discussion above is meant to give the reader a sound scepticism with respect to the analysis in the rest of the dissertation. Having said that, the focus is shifted to a summary of some of the literature on the modelling of sustainability, abatement, capital accumulation and population growth, human capital and finally the strategic aspects of international cooperation.

2.3.2 Sustainability

This section describes the concept of sustainability and operational definitions of this concept. Next, the specification of the assimilation function of nature is discussed.

The Concept of Sustainability

As the term 'sustainability' is more and more in vogue, the ambiguity over its meaning is ever increasing. Pezzey (1989) lists some 60 definitions of sustainability, sustainable growth and sustainable development that exist in the literature. One of the most commonly used definitions of sustainability in environmental economics is that of 'the maintenance of the effective resource base'. In order to operationalise this concept, it is crucial to know the particular resource(s) of interest. Recalling the subdivision of resources into renewable, exhaustible and environmental resources given in Chapter 2, sustainability can be defined for each of them respectively as (see Barbier and Markandya, 1989, p.2,3):

- utilising *renewable resources* at rates less than or equal to the natural or managed rates of regeneration;

- optimising the efficiency with which *exhaustible resources* are used, which is determined, among other things, by the rate at which renewable resources can be substituted for exhaustibles and by technical progress;

- generating waste and pollution (i.e. *the environmental resources*) at rates less than or equal to the rates at which they can be absorbed by the assimilative capacity of the environment

Note that sustainability should be distinguished from sustainable development. The poldering of lakes in the Amsterdam region in the fifteenth century was not sustainable for the renewable resources in these lakes (fish and plants). However, it is difficult to claim that the poldering was therefore bad for the sustainable development of the Netherlands. Hence, one has to be careful to really use the sustainability criterium on a species level[42]. However, for the Greenhouse Effect and for many other problems of the environmental resource type, sustainability is a *sine qua non*. The reason is that production will go to zero for too elevated levels of 'pollution', as was assumed in the last section. Therefore, in the absense of sustainability of the resource use, sustainable development is not possible[43].

---

[42]At a total stock of natural resources level, however, another problem is on the lurk: that is how to weigh the different stocks (see Pezzey, op. cit., p.15).

[43]In the functional form chosen for the Greenhouse damage function, $f(P)$, the economy collapses when P approaches a threshold level $\bar{P}$. At that point production and consumption are zero.

In Nordhaus' (1982) model above sustainability implies that the long term flow of GHGs into the atmosphere (i.e. γF^{∞})[44] should be (less than or) equal to the rate that can be absorbed by the assimilative capacity of the environment (i.e. αP^{∞}). This amounts to saying that sustainability constrains $\dot{P} = \gamma F - \alpha P$ to be (less than or) equal to zero in the long run. This is all rather straightforward, given the simplicity of the formulation of the assimilation function of nature α. Below, attention is focused on more realistic functional forms of the natural regeneration process.

The Assimilative Capacity of Nature

As discussed in Section 2.2 above, relatively little is known about the assimilation process of Greenhouse gases in the environment. Many authors, having stressed that biologists could not supply them with nice quantitative assimilation functions, take the linear specification as a computationally convenient proxy[45]. Counter-examples are, among others, Forster (1975), Dasgupta (1982), Barbier & Markandya (1989) and Pethig (1990), stressing the multiple equilibria problem that may result.

Dasgupta (1982) points out that there may be three phases in the assimilation process. Firstly, at low 'pollution' levels, assimilation increases with pollution. Next, 'pollution' becomes constant, followed subsequently by a decreasing trend leading to a collapse of the eco-system.

Pethig (1990) chooses a specific general form for the assimilation function. At low levels of pollution, assimilation has increasing returns to scale, followed by decreasing returns for higher levels of pollution. After a certain point, the total assimilation declines till there is no regenerative capacity left[46,47]. Forster (1975) has a similar specification, but concentrates on the concave part of the assimilation function.

---

[44] The superscript ∞ denotes the steady state value.

[45] Another convenient approximation is to take the assimilation function as a constant term, as does Mäler (1989).

[46] Pethig (1990) uses environmental quality, Q, instead of pollution as the independent variable in $A(.)$. With Q_v as the environmental quality without pollution (the 'virgin state') and defining $P \equiv Q_v - Q$, the description given here becomes identical to Pethig's formulation.

[47] Formally, defining $A(P)$ has the assimilation function, Pethig describes the functional form as follows. There is a $\tilde{P} > 0$, such that $A(\tilde{P}) = 0$ and $A(P) = 0$ for all $P \geq \tilde{P}$. Moreover, A is continuously differentiable on $[0, \infty)$, is positively valued on $\langle 0, \tilde{P} \rangle$, and there is an inflection point, $P_f \in \langle 0, \tilde{P} \rangle$, such that $A_{PP} > 0$ for $P \in \langle P_f, \tilde{P} \rangle$ and $A_{P,P} < 0$ for $P \in \langle 0, P_f \rangle$

Barbier & Markandya (1989) have developed a model where the assimilation function is not defined separately, but together with emissions. This is achieved by replacing Nordhaus' (1982) equation of motion $\dot{P} = F - \alpha P$ with:

$$\dot{P} = \begin{cases} h(F,P) & \text{for } P \leq \tilde{P} \\ \ll 0 & \text{for } P > \tilde{P} \end{cases} \tag{2.7}$$

Barbier & Markandya (1989)[48] allow for a collapse of the ecosystem where positive feedbacks that deteriorate the environment take place. Formally: $h(F,P) \ll 0$ (even for $F = 0$!).

How the assimilative capacity of the environment is modeled depends very much on the ecosystem that is described and is in principle an empirical matter. Unfortunately, biologists stress that very little is known about the true process of assimilation[49]. Therefore, in Chapter 3, different specifications of the assimilation function will be considered. This gives an idea of the effects that slight variations in the functional form may have on the steady state levels of pollution and consumption.

2.3.3   Emission Reduction with Fixed Capital Stock

In environmental policy, at least four forms of emission reduction are distinguished:

- decrease of production;

- abatement (e.g. end-of-pipe);

- recycling of materials;

- process-integrated changes.

In the Nordhaus (1982) model discussed above[50], the only way to reduce emissions is by producing less. In fact, recent decreases in pollution in Central & Eastern Europe are mainly due to a close-down of production units (often not for environmental reasons).

---

[48]Barbier & Markandya (1989) take a specification with consumption C instead of F. Here, F is taken to be comparable with Nordhaus (1982).

[49]The only certainty is that in reality, it is far more complex than can ever be captured in a simple function αP or $h(C,P)$.

[50]In this model, consumption equals production ($C = f(F,P)$) and the pollution stock changes over time according to $\dot{P} = \gamma F - \alpha P$.

Instead of producing less, it is often cost-efficient to use abatement techniques in order to reduce pollution. Examples include water purification, flue gas desulphurisation and many more. Reforestation can actually, in the case of the Greenhouse Effect, be seen as a special form of abatement[51]. Abatement can either be applied at the source, with for instance the use of filters (end-of-pipe techniques) or at a later stage, after the discharge into the environment has taken place (water purification plants).

Recycling goes theoretically one step further than abatement. The idea is that, as with abatement, polluting material is captured so that it does not disperse into the environment, but, this material is now re-used as a productive input after it has been collected.

Process-integrated changes or process innovations imply that there is a shift away from pollution inputs and/or processes in the production. In order to materialise such a shift, a second input needs to be introduced in the production process. Straightforward candidates are capital and labour. An example is the use of pesticides. Labour (and/or capital) can be used as an input for the production of pesticides and also as an input in the agricultural sector. Measures to reduce the use of pesticides can lead to a shift of labour (and/or capital) to the agricultural sector.

Each of these forms of emission reduction, as an alternative to decreases in production, is summarised in Table 2.3 and will be discussed below for fixed capital and labour. In the next section, capital accumulation and population growth will be considered.

Abatement

Emission reduction in the form of abatement can be added to the model of Nordhaus (1982), by taking:

$$C = f(F, P) - k(A) \tag{2.8}$$
$$\dot{P} = \gamma(F - A) - \alpha P \tag{2.9}$$

where A is abatement and $k(A)$ is the net production cost of abatement. Implicitly, it is assumed that the production of abatement equipment is itself not polluting. Note that only the left-over emissions $(F - A)$ contribute to environmental degra-

---

[51]By increasing forestation, emitted carbon dioxide can be recaptured in the same way that discharged water pollution is separated by filters.

Emission Reduction in Models without Capital Accumulation			
decrease in production	abatement	recycling	process-integrated changes
Nordhaus (1982)	Plourde (1972) Forster (1977) Dasgupta (1982) B&M (1989) VDP&W (1991-M2)	Smith (1972)	KSZ (1971-M2)

Table 2.3: *Summary of Control Models without Capital Accumulation given below (B&M stands for Barbier & Markandya; VDP&W denotes for Van der Ploeg & Withagen; KSZ stands for Keeler, Spence & Zeckhauser; M is 'model')*

dation. Optimising welfare $(U(C))$ over time (as in Equation 2.4) with respect to F and additionally A gives the optimal trajectory of the relevant variables[52].

Forster (1977) analyses a model rather similar to the prototype model discussed above. There are three differences. First, Forster (1977) concentrates on the dynamic trade-off between present consumption and future pollution, using a social welfare function $U(C,P)$[53]. This means, in the terminology of Pezzey (1989) that pollution has an 'amenity effect' rather than a 'productive effect'. A further difference is that output is assumed to be fixed. This output, $\bar{\phi}$, can be used for consumption goods, C, and for abatement, A. Formally: $\bar{\phi} = C - A$. Finally, consumption is taken to contribute with rate $g(C)$ to environmental degradation. Pollution is cleaned up as a function $h(.)$ of the amount of abatement expenditures:

---

[52]Compared to the case without abatement possibilities, the introduction of A will, *ceteris paribus*, lead to a higher level of F and hence of output $f(.)$. However, if abatement is relatively too expensive, it may be optimal to set $A = 0$. If, on the other hand, abatement is relatively cheap, the actual energy use (and hence the emission level) could be as high as its level in the absence of environmental degradation ($\bar{F}$). However, assuming that the marginal costs of the first unit of output reduction and abatement are zero, the optimum is generally an internal solution in which $A > 0$ and $f(F) < f(\bar{F})$

[53]Instead of a social welfare function, Dasgupta (1982) uses a cost-benefit framework of the social benefits of output and the costs of environmental degradation and investment in abatement. Again other models compare the benefits of decreases in environmental degradation with abatement costs. A comparison of the different objective functionals shows that it all boils down to the same trade-off of consumption now versus pollution in future.

$h(A)$. It is assumed that $g(.)$ is convex and $h(.)$ is concave. The stock of pollution will increase with $g(C)$ and decrease with $h(A)$ and assimilation αP:

$$\dot{P} = g(C) - h(A) - \alpha P \qquad \text{with} P > 0 \qquad (2.10)$$

$$\bar{\phi} = C - A \qquad (2.11)$$

The direct link between emissions and consumption implies that it is not a polluting input (as in the prototype model above) that leads to environmental degradation (as in Nordhaus, 1982), but the output (or consumption) as such. In Section 2.3.1 above these two different specifications have been referred to as 'polluting input models' and 'polluting output models' respectively.

Van der Ploeg & Withagen (1991; Section 2) discuss a very similar model[54]. Here however, total output, instead of consumption, contributes to pollution. This means that the production of abatement services is also assumed to be polluting. A further distinction is that abatement is modeled as a way to enhance natural assimilation rather than to reduce emissions/discharges. Formally:

$$\dot{P} = \gamma Y - \alpha(A)P \qquad (2.12)$$

where Y denotes production. The specification of the assimilation function could be an apt description in the case of water pollution, where the natural clean-up can be stimulated by artificial aeration of water bodies with low oxygen content (Dasgupta, 1982, p. 157).

A more general model with basically the same structure is by Dasgupta (1982). The evolution of the pollution stock is described by:

$$\dot{P} = \epsilon(I)Y_t + J(G_t, P_t) \qquad I, Y_t, P_t \geq 0 \qquad (2.13)$$

where: $\epsilon(.)$ is the emission-output ratio; Y_t is the aggregate production at time t; G_t is the human effort to enlarge the assimilative capacity of nature at time t and I is the capitalised value of pollution abatement costs. Net benefit is optimised with respect to all three controls Y_t, G_t and I.

The main extension of this model, compared to the ones discussed above, is the introduction of abatement investment, for instance in end-of-pipe technology. This investment will influence the emissions per unit of production. It is assumed

---

[54]This model is based on Dasgupta (1982), discussed below. For the sake of argument, Van der Ploeg & Withagen (1991; Section 2) will be discussed first.

that the emission-output ratio $\epsilon(I)$ is decreasing in investment. Furthermore, $\epsilon(I)$ is assumed to be convex. In the absence of abatement efforts, $\epsilon(0)$ is assumed to be the emission-output ratio corresponding to the 'cheapest' production technology[55].

Note also that the function, $J(G_t, P_t)$, is more general than the formulation $\alpha(A)P$ of Van der Ploeg & Withagen (1991). With this broader specification, reforestation in order to combat the Greenhouse Effect and eco-engineering techniques like shooting ozone into the stratosphere to stop ozone depletion can be modeled[56].

A very general formulation of resource problems is given by Barbier & Markandya (1989). The evolution of the stock of pollution over time is modeled as:

$$\dot{P} = h(C, P) \tag{2.14}$$

Barbier & Markandya (1989) assume $h(C, P)$ to be increasing and convex in both C and P. The broad formulation $h(C, P)$ for net environmental degradation allows Barbier & Markandya to incorporate both waste, pollution and renewable resources in the model[57,58].

A quite different model is by Plourde (1972), who explicitly introduces input factors in the production process that can be shifted from the production of consumer goods to abatement services[59]. Assume that the total amount of capital is fixed, $(K = \bar{K})$. It can be used either to produce a consumer good $f(K_1)$ or to abate pollution $g(K_2)$. Both $f(.)$ and $g(.)$ are assumed to be concave.

Note that a shift from K_1 to K_2 has two distinct positive effects on the flow of pollution, first because of the lower level of pollution through $f(K_1)$ and secondly because of higher abatement activities. This results from the fact that the production of abatement technology is not polluting.

---

[55]Dasgupta (1982) is rather vague about I. The fact that a time-index is left out of the formulation implies that I is fixed. This means, however, that a new stock variable is introduced via the back-door. Therefore, I would prefer a formulation with either an annual flow of abatement services I_t or a formulation with $\epsilon(K_t)$, where K_t is the stock of abatement technology. This issue is further analysed in the next section.

[56]Both are difficult to model in the formulation of Van der Ploeg & Withagen (1991), as these techniques do not really increase the rate of assimilation, but rather change the stock of pollution directly

[57]Barbier & Markandya also incorporate abundant exhaustible resources in the model.

[58]As explained in Section 2.3.2 above, the equation of motion is in fact more general. Above a threshold level $\tilde{P}$, it is assumed that $\dot{P} \ll 0$

[59]Plourde (1972) uses labour instead of capital as input. Besides, Plourde deals with waste, though here the air-pollution terminology (emissions, etc.) will be retained here.

The evolution of the stock of pollution (P) depends on the pollution associated with the production of the consumption good, $\gamma f(K_1)$, on the production of the abatement industry, $g(K_2)$ and on the assimilative capacity of the environment, αP:

$$\dot{P} = \gamma f(K_1) - g(K_2) - \alpha P \qquad (2.15)$$
$$\bar{K} = K_1 + K_2 \qquad (2.16)$$

Note that the specification with 'net-pollution' $\gamma f(K_1) - g(K_2)$, as stressed before, is more appropriate for water pollution problems where actual clean-up is possible then for air pollution problems where abatement is a portion of the emission level (see Dasgupta, 1982)[60]. Plourde assumes that there exists a social welfare function, $U(C, P)$.

Optimal solutions can be achieved either through a *Central Planning Bureau*[61], dictating the whole economic process or through a market economy[62]. It is straight-

---

[60]This is true unless clean-up possibilities exist in the form of, for instance, reforestation in order to combat the Greenhouse Effect (see also Dasgupta (1982)).

[61]The intertemporal optimisation of the Central Planning Bureau with C, K_1 and K_2 as control variables can be formalised as:

$$\max_{C, K_1, K_2} \int_0^\infty e^{-\rho t} U(C, P) \, dt \qquad \rho > 0 \qquad (2.17)$$

$$\text{s.t.} \qquad \dot{P} = \gamma f(K_1) - g(K_2) - \alpha P \qquad (2.18)$$
$$K_1 + K_2 = \bar{K} \qquad (2.19)$$
$$f(K_1) \geq C \qquad (2.20)$$

[62]In order to decentralise the model of Plourde (1972), an additional representative producer must be introduced compared to the Nordhaus (1982) model in Section 2.3.1. Hence, the economy is supposed to consist of two representative producers, one for the consumption good and one for the abatement industry. The actors of this economy are:

The Representative consumer who maximises utility $U(C, P)$ over time subject to: $C = \pi_1 + \pi_2 + rK_1 + rK_2 + T$;

The Government with sets pollution taxes τ, given his budget constraint $T = \tau\{\gamma f(K_1) - g(K_2)\}$;

The Representative firm producing the consumer good who maximises profits $\pi_1 = \lambda f(K_1) - rK_1 - \tau\gamma f(K_1)$ with respect to K_1;

The Representative firm producing the abatement services who maximises profits $\pi_2 = \tau g(K_2) - rK_2$ with respect to K_2.

forward to show that if, in the decentralised economy, the government sets taxes, corresponding to the shadow price of pollution in utility terms, then this economy yields exactly the same results as the one ruled by a central planning bureau. Hence, the government with perfect information can control emissions optimally by setting the externality tax equal to the shadow costs of pollution[63].

This description of the model of Plourde (1972) ends the discussion on pollution control through abatement in economies without capital accumulation or population growth. The analysis shifts now to other forms of emission reduction (recycling and process-integrated changes), followed by a general discussion on the main characteristics of emission reduction models.

Recycling

Along the lines of Plourde (1972), but with three instead of two sectors in the economy is the recycle model of Smith (1972). It is assumed that the economy produces q_1 units of a commodity together with the same amount of waste. This makes sense, for instance, if the commodity is milk and the waste is the amount of bottles. Instead of producing bottles, they can also be recycled. Assume that the amount q_2 will, in fact, be recycled, leaving the amount of waste that will be disposed of at $q_3 = q_1 - q_2$. Therefore the stock of pollution changes over time according to:

$$\dot{P} = q_1 - q_2 - \alpha P \qquad (2.21)$$

It is assumed that capital[64] is the only input in the production of the commodity, of the waste and of the recycling: $q_i = f_i(K_i)$ for i = 1,2,3. Each $f_i(K_i)$ is concave and $f_i' > 0, f_i(0) = 0$. The total amount of capital is fixed, so that $\bar{K} = K_1 + K_2 + K_3$.

The social welfare function consists of utility from the commodity, disutility from recycling, and disutility from the stock of waste: $U(q_1, q_2, P)$. Optimising this

---

[63]Note that, in the absence of taxation, the representative producer of the consumer good will maximise $\pi_1 = \lambda f(K_1) - rK_1$ and the producer of abatement services will cease to exist. This means that the market rate of return will be $r = \lambda f_{K_1}(\bar{K})$ and hence the observed rate of return is higher than the socially optimal one, with corresponding higher levels of pollution. Hence, the excessive level of pollution due to the distortions of imperfect markets can be corrected by emission taxes, where the appropriate discount rate is the socially optimal one and not the observed rate of return before introduction of the emission charge. This is the point made in Section 2.3.1 (see also Solow (1973) and Mäler (1974)).

[64]Smith (1972) takes labour instead of capital.

function over time generally gives positive levels of q_i for all i, though, boundary solutions cannot be excluded.

Process-Integrated Change

Theoretical models on emission reduction, like the ones described above, typically assume that economic incentive-based policy instruments (as well as other instruments) primarily affect the level of abatement. In practice, though, government environmental policy induces firms to re-analyse the entire production process. Ideally firms critically screen all the waste and pollution streams originating from all inputs and throughputs and from all aspects of output (including packing material, etc.). Charges on specific elements of the production tend to encourage a shift away from these elements towards less polluting elements. An example of how this can be modeled is Keeler, Spence & Zeckhauser (1971; Model 2)[65].

In their specification, capital[66] can be allocated to the manufacturing of the consumption good and the production of a pollutant. At the same time, the pollutant, serving as an intermediate good, is the other input in the production function:

$$C = f\{K_1, j(\bar{K} - K_1)\} = f(K_1)$$

where: $j(\bar{K} - K_1)$ is the primary flow of the produced pollutant, so that pollution changes over time as $\dot{P} = j(\bar{K} - K_1) - \alpha P$[67]. The reasoning behind the model can best be understood by thinking of agricultural production using pesticides, such as D.D.T., the production of which uses part of the capital stock. The use of pesticides enlarges the output of the agricultural production, but at the same time the stock of the pesticide has a negative influence on social welfare.

This model of Keeler, Spence & Zeckhauser serves as an example of a situation where pollution is produced as an intermediate good and that it is controlled through a shift in the choice of production process. This can be seen as an extension of the 'polluting input' model of Nordhaus (1982). In both specifications, it is an input (or intermediate good) and not the output that causes the pollution.

---

[65]The famous two state variable model in the same paper (Keeler, Spence & Zeckhauser (1971); model 1) will be described in the next section.

[66]Keeler, Spence & Zeckhauser take labour as the production factor. For reasons of comparison, this is changed here into capital.

[67]In fact, Keeler, Spence & Zeckhauser (1971) take $\dot{P} = a(\bar{K} - K_1) - \alpha P$, where $a(.)$ differs from $j(.)$ in the production function.

Up to now, the specification of different models has been described, concentrating on the production process and on the change in pollution over time. It is important to stress that there is no such thing as a bad or a good model to describe pollution, emissions, abatement, etc. At the same time it is crucial to see in which situations the different kinds of model can be used. For instance, the assimilation function by Van der Ploeg & Withagen (1991) could be an appropriate formalisation of artificial aeration of water bodies. The idea of an emission-output ratio $\epsilon(.)$ of Dasgupta (1982) might capture different abatement options[68] for sulphur dioxide emissions from power plants. Similarly, the specification of net pollution as $\gamma f(K_1) - g(K_2)$ by Plourde (1972) may be a good description of abatement of water pollution.

Some of the elements analysed here that seem to be useful for a specification of emission reduction possibilities with respect to the Greenhouse Effect will be adopted in Chapter 3.

Finally, the optimisation results of the models discussed above will be briefly summarised. The standard result is a unique saddlepoint equilibrium with positive levels of pollution and abatement (and/or other forms of emission reduction) brought about by polluter charges (e.g. Keeler, Spence & Zeckhauser (1971; model 2); Forster (1977), Smith (1972) and van der Ploeg & Withagen (1991)).

The result that positive levels of pollution are optimal hinges on the assumption that marginal damage costs are negligible for very low levels of pollution (formally: $U'_P(C,0) = 0$ in the utility function $U(C,P)$). This is analysed in a critical note by Forster (1972) regarding the conclusion of Keeler, Spence & Zeckhauser (1971) that non-zero levels of pollution are optimal. Forster (1972) shows that by changing their model slightly, using $U'_P(C,0) = -b$ with $b > 0$, zero pollution can be optimal. This could be especially realistic for very toxic substances with slow natural decay.

Positive levels of abatement (or recycling) are optimal as a result of the assumption with respect to the marginal costs for abatement and environmental degradation. For instance, if people do not care very much for the environment and recycling is expensive, zero recycling can be optimal. This is analysed formally by Smith (1972, p. 607)[69].

Uniqueness and stability depend on the concavity/convexity assumptions of the

---

[68]Possible options are limestone injection and other forms of combustion modification, flue gas desulphurisation and regenerative processes.

[69]Similarly, 100 % recycling or abatement could be optimal if the amenity value of the environment is high and recycling, even for high percentages, is low.

model. Forster (1975) shows this by assuming a non-linear assimilation function $\alpha(P)P$ instead of αP (as in Forster, 1977 and other papers). If $\alpha(P)P$ is concave for positive values and zero elsewhere (and hence not concave over the whole domain of P), multiple stable equilibria cannot be excluded[70].

Finally, the use of emission charges (Pigouvian taxes) or other forms of market-based instruments for the design of environmental policy in order to secure a social optimum hinges on the assumption that there are no 'fundamental non-convexities' of the production set at the relevant part of the domain (Starrett, 1972; Pethig, 1990, p.4).

2.3.4 Emission Reduction and Capital Accumulation

The dynamics in the models described above originate from the evolution of the stock of pollution over time. In this section, the issues of population growth and capital accumulation are described adding to the complexity of the dynamics. Population growth is only briefly described and the main body of this section is devoted to capital accumulation.

Even if environmental resource use per capita does not increase, *population growth* can lead to considerable stress on the environment. Therefore, the issue of population growth seems to be extremely important for problems, such as the Greenhouse Effect.

In the writings of Malthus and more recently in the Report of the Club of Rome, exponential population growth with finite natural resources is the key issue. Solow argues, however, that in a model with "finite natural resources, its seems ridiculous to hold to the convention of exponentially growing population. We all know that population cannot grow forever if only for square-footage reasons" (Solow, 1974, p. 36). Therefore, Solow assumes zero population growth in his long term models[71].

Mäler (1974) and Dasgupta & Heal (1974), on the other hand, argue that ideally population policy should be included in natural resource models[72].

---

[70]Barbier & Markandya (1989) allow the assimilation function to be negative, so that it is concave over the whole domain of P. They analyse with this specification the case with one stable and one unstable equilibrium.

[71]Assuming a logistic growth curve of population, Solow's arguments makes sense. However, it can still be that in the medium-term population growth can well be approximated by an exponential function.

[72]This could be modeled in a logistic growth setting where the ultimate size of the population or the speed to arrive at this size is endogenous.

D'Arge & Kogiku (1973) model a time dependent utility function, where nature has no assimilative capacity and only a fixed proportion of waste is recycled. Given a maximum viable level of waste and fixed population growth, the aim of the paper is to select the admissible control output so that a maximum integral of discounted utility is achieved over the interval $0 \leq t \leq \tau$, where τ is the endogenous end-point.

Some other economists have tried to model population growth in an endogenous way, depending on the level of technical progress (Becker, Murphy & Tamura (1990)) or depending on the state of the environment (Nerlove & Meyer (1991)). Notwithstanding its obvious relevance, the population issue will not be elaborated later on.

The rest of this section focuses on *capital accumulation* in Ramsey type models with an environmental stock variable. Modeling of capital is of crucial importance, as it highlights the role of investment decisions brought about by incentives from the market and from the public sector. Again, a distinction is made between models of abatement, recycling and process-integrated change. Table 2.4 gives a brief summary of the existing models that will be discussed.

Emission Reduction in Models with Capital Accumulation			
decrease in production	*abatement*	*recycling*	*process-integrated changes*
	KSZ (1971-M1)	*Mäler (1974-M1)*	*Brock (1977)*
	Mäler (1974-M1)	*Smith (1972)*	*T&K (1990)*
	Musu (1990, 1992)		*Pethig (1990)*
	VDP&W (1991-M2)		

Table 2.4: *Summary of Control Models with Capital Accumulation given below (VDP&W stands for Van der Ploeg & Withagen; KSZ denotes Keeler, Spence & Zeckhauser; T&K stands for Tahvonen & Kuuluvainen; M is 'Model')*

Abatement

In their seminal paper in 1971, Keeler, Spence & Zeckhauser describe a model of abatement where pollution is a by-product of production. In the terminology introduced above, this is referred to an a 'polluting output' model. There is only one capital good, that is used for the production of output. This output can be used for consumption, capital investment and abatement. Pollution is a fixed fraction of output. A central planner is assumed to maximise intertemporally a utility function, $U(C, P)$, which includes consumption and pollution. The proportions of output spent on consumption and abatement are the control variables in the model. Hence:

$$C = \gamma f(K) \tag{2.22}$$

$$\dot{P} = (1 - \beta d)f(K) - \alpha P \tag{2.23}$$

$$\dot{K} = (1 - \gamma - \beta)f(K) - \delta K \tag{2.24}$$

where γ and β are the proportions of production used for consumption and abatement purposes respectively and d is the productivity of the abatement services.

Note that $(1 - \beta d)$ corresponds to the emission-output ratio of Dasgupta (1982). Keeler, Spence & Zeckhauser (1971) assume that this ratio relates linearly way to the fraction of production spent on abatement. This implies that it is possible to shift instantaneously from low abatement to high abatement. Dasgupta has, in my opinion, a more realistic assumption that the emission ratio depends on the accumulated investments in abatement. This approach is also taken by Musu (1992) in a model discussed below[73].

Van der Ploeg & Withagen (1991; Section 7) use a very similar framework where abatement enhances the assimilative capacity of nature and does not lower the emission-output ratio[74]. Hence:

$$\dot{P} = af(K) - \sigma(A)P \tag{2.25}$$

$$\dot{K} = f(K) - \delta K - C - A \tag{2.26}$$

where δ is depreciation and a is the fixed emission-output ratio. The rest of the symbols are straightforward (cf. Equation 2.12).

Mäler (1974, Paradigm I) uses the same type of model, with the additional feature that the production of output is brought about by extraction of a natural

---

[73]It is assumed that $U_C(0, P) = \infty$, hence it follows that there are are at most two solutions: one is the boundary solution with $\beta = 0$ (no abatement) and the other is the internal solution with $\gamma > 0$, $\beta > 0$ and $0 < 1 - \beta - \gamma < 1$. Keeler, Spence & Zeckhauser (1971) refer to these cases as the 'murky age' and the 'golden age' respectively.

[74]In fact, Van der Ploeg & Withagen (1991, p. 223) claim that Keeler, Spence & Zeckhauser (1971) do only model cleaner technology and do not use abatement. I do not agree with this claim. Instead, I think Keeler, Spence & Zeckhauser (1971) do use abatement leading to a lower fraction of pollution emitted, as with putting filters on smoke-stacks.

resource[75]. The total amount of the resource is limited. This is an interesting example of modelling economic policy in the presence of both environmental and exhaustible resources. A real life example is fossil fuel use, that is limited in quantity and which is polluting in its use[76].

In the models discussed up to now, abatement is seen as a flow (a part of output). Musu (1990), on the other hand, introduces abatement as a stock of capital. In his vision, abatement capital, K_2, and productive capital, K_1 form together the composite capital good, K[77]. Output is produced as a composite commodity that can either be used for consumptive purposes or for the accumulation of capital (both K_1 and K_2). Production is polluting, but the environment can be cleaned up with the help of abatement capital, K_2. A social planner is assumed to maximise welfare, $U(C, P)$ over time[78], using consumption and abatement capital as the two control variables[79], subject to:

$$\dot{P} = af(K_1) - g(K_2) - \alpha P \tag{2.27}$$

$$\dot{K} = f(K_1) - C - \delta K \tag{2.28}$$

$$K = K_1 + K_2 \tag{2.29}$$

where g(.) is the abatement function and a is the emission-output coefficient.

In order to allocate K_1 and K_2 optimally, "the social value of the marginal productivity of capital allocated to the first sector, net of the social value of the detrimental externality due to pollution emission, must be equal to the social value of the marginal productivity of capital in the abatement sector" (Musu, 1990, p. 7)[80]. He shows that for sufficiently low values of social time preference, the steady state is a saddle point. In a more recent paper, Musu (1992), reformulates this

---

[75]Production, $f(K)$, measures both the volume of output and the rate at which the natural non-renewable resources are exploited (Mäler, 1974, p.63).

[76]The model is worked out in a finite horizon time framework. Mäler sets the discount rate to zero, as a positive rate together with scarce resources prevents feasible steady states from being optimal.

[77]Note that this is basically the same formulation as Plourde (1972), discussed above. In that model composite capital was kept constant, $\bar{K}$, as expressed in Equation 2.16, whereas here, K accumulates over time.

[78]In fact, Musu describes his model in terms of environmental quality, Q, and not in terms of pollution. As noted in Section 2.3.1 above, the two formulations are equivalent.

[79]This gives rise to an infinite time free right-hand endpoint optimal control problem with pollution, P, and capital, K, as the two state variables.

[80]Musu (1990) describes the dynamics of the model in great length, linearising the system of dynamic equations in order to derive conditions for local stability

model in such a way that 'net pollution' is not any more the difference between two functions of the capital stock, $f(K_1) - g(K_2)$, but a product of two functions of capital stocks, $h(K_2)g(K_1)$. In this formulation, $h(K_2)$ is an emission-output ratio that decreases with investment in abatement technology.

Luptáčik & Schubert (1982) introduce a three-state variable model, similar to Musu (1990). Their model has, like Musu's formulation, two capital stocks (K_1 and K_2). The difference is that Luptáčik & Schubert (1982) introduce separate capital accumulation functions for the two capital stocks[81]. The model by Luptáčik & Schubert (1982) can be summarised as:

$$\dot{P} = f(K_1, P) - h(K_2, P) - \alpha P \tag{2.30}$$

$$\dot{K_1} = I_1 \tag{2.31}$$

$$\dot{K_2} = I_2 \tag{2.32}$$

$$f(K_1, P) = C + I_1 + I_2 \tag{2.33}$$

Note that there is no explicit capital depreciation. In order to analyse the stability properties of the model, the authors reduce the dynamic system to a two-state variable model. This is achieved by assuming that the production function, $f(.)$ and the abatement function, $h(.)$ are linearly homogenous. Furthermore, the utility function is simplified to $U = U(C/P)$. The two remaining state equations are K_1/P and K_2/P. The resulting two-state variable problem can be shown to be locally asymptotically stable (see Luptáčik & Schubert, op. cit., p.241).

Recycling

In order to describe recycling, Mäler (1974, Paradigm 4) adapts his model presented his Section 2.3.4 to allow for recovery of waste streams. In this framework, K_1 denotes the capital stock employed in the production of the composite good, $f(K_1)$, and K_2 is the capital stock allocated to recovery. The sum, $K = K_1 + K_2$ is the total capital stock in the economy. Output, $f(K_1)$, can be used for capital accumulation, consumption and for investment in recovery technology, I_2. The production side of the model can be formalised as follows:

$$\dot{K_1} = f(K_1) - C - \delta K_1 - I_2 \tag{2.34}$$

---

[81] Another difference is that the production functions for consumption, $f(.)$, and for abatement, $g(.)$, depend on the capital input as well as on the stock of pollution. This means, that pollution has both an amenity effect and a productive effect (see Section 2.3.1).

$$\dot{K_2} = I_2 - \delta K_2 \tag{2.35}$$

$$K_2 \geq g(v,x) \tag{2.36}$$

$$x = C + \delta(K_1 + K_2) \tag{2.37}$$

$$x = z + v \tag{2.38}$$

where x is the total flow of residuals, v is the flow of raw materials recovered, z is the net flow of residuals discharged into the environment and $g(v,x)$ is the capital requirement function.

Note that Mäler assumes that residuals are a by-product of capital depreciation and of consumption and not of production. Further, as in his model discussed above, the total amount of resources is limited. Therefore, investment in recovery technology have the double beneficial effect of postponing the depletion of the exhaustible resource and saving the environment. Government policy directed to high resource prices and subsidies on investment in recovery technology can induce such beneficial shifts.

Process-Integrated Change

Changes in the production process away from polluting inputs towards less polluting ones have been modeled by Brock (1977). The production function has inputs K (capital) and F (energy, pesticide, etc.) and the focus is on the substitution possibilities between these two inputs. Examples in the case of the Greenhouse Effect are investment in insulation, energy saving technology, biomass technology, windmill installations, etc. The model of Brock is elegant in its simplicity. A social planner is assumed to maximise welfare, $U(C,P)$, over time, subject to

$$\dot{P} = F - \alpha P \tag{2.39}$$

$$\dot{K} = f(K,F) - C \tag{2.40}$$

Note that in the terminology introduced in Section 2.3.1 this is an 'polluting input model'. Brock (1977) works out this model in its decentralised form.

Tahvonen & Kuuluvainen (1990; model I) analyse the requirements for stability and uniqueness of Brock's (1977) model[82] Moreover, they focus on the substitution possibilities between K and F by defining $\theta \equiv K/F$ and by assuming that $f(.)$ is

---

[82]For instance, a steady state may fail to exist for too large values of time preference, too low rates of decay of the stock of pollution and for a too low elasticity of substitution. As is also shown in Forster (1977), non-uniqueness cannot be ruled out if $U_{C,P}$ is allowed to be positive.

homogeneous of degree one. This means that the steady state capital-energy ratio (and hence the capital-emission ratio) changes according to:

$$\dot{\theta}/\theta = \sigma\phi(\theta)/\theta \qquad (2.41)$$

where $\phi(\theta) = f(K,F)/F$ and σ is the elasticity of substitution between K and F. Modeling the capital-energy ratio was central in the discussion on energy scarcity in the seventies (e.g. Dasgupta & Heal (1974), Stiglitz (1974) and others). The difference betwee that discussion and the one of Tahvonen & Kuuluvainen (1990) is that energy scarcity means that θ has to grow fast enough to maintain a non-declining production. In the environmental case, θ should growth fast enough to maintain environmental quality[83].

Pethig (1990) describes a model that is quite similar to the model of Brock (1977) and Tahvonen & Kuuluvainen (1990). The main difference is that the production function not only depends on capital, K, and on the flow of pollutants, F, but also on the stock of pollutants, P. This means that the stock of pollution has both an amenity effect and a productive effect:

$$f = f(K,F,P) \qquad (2.42)$$

with $f'(K) > 0$, $f'(F) > 0$ and $f'(P) < 0$. Note that as an externality, the pollutant creates a 'fundamental non-convexity' of the production set. Pethig (1990) assumes though that on the relevant part of the domain, $f(.)$ will be concave.

The environmental quality changes with the discharges F, therefore, Pethig's model is an example of what is referred to as a 'polluting input model'. In Pethig's words: "Observe that in $f(.)$ the emission is treated as if it were an input even though it is clearly an undesired output. But the input interpretation is both appropriate and convenient, because in addition to being an output, the emission constitutes the industry's *demand for a productive factor*, namely the *waste assimilation services* of the environment" (Pethig, op. cit., p.4)[84].

In this section, different types of models of emission reduction with capital accumulation have been discussed. Capital can both be used for consumption and

---

[83]Tahvonen & Kuuluvainen, in the same paper describe a model with has additionally a renewable resource. The growth of this resource is negatively affected by the presence of pollution. Modelling this leads to a three-state variable model.

[84]Pethig's main aim is the description of learning and uncertainty about the resource characteristic. To this aim a quasi-option value is introduced.

investment purposes and for limiting environmental degradation. Tahvonen & Ku-uluvainen (1990) and Van der Ploeg & Withagen (1991) show that the equilibrium stock of capital is smaller in the case with environmental concern than in the case that the social planner does not care about the environment (co-state variable of pollution is zero).

For the rest, the same conclusions hold as for the models without capital accu-mulation, though the concavity/convexity requirements for stability and uniqueness are much more stringent. Often, only local stability can be guaranteed. Also, the rate of time preference needs to be sufficiently low. At the end, however, it is an empirical question whether these assumptions are realistic.

As a next step, investment in knowledge is introduced which will lead to en-dogenous human capital accumulation.

2.3.5   Emission Reduction and Human Capital

In the last section, physical capital accumulation was described. Now we focus on the question of technological progress that can make capital more productive. Specifically, we are interested in models with deliberate investment by the pri-vate sector in knowledge/skills in order to increase its productivity and hence to produce the same amount of output with less input. The importance of the endo-geneity of technical change was stressed forcefully by Mäler (1974) and Solow (see Section 2.3.1). For instance, Mäler (1974) argues that "technological progress is not something that comes to the economy independent of the actions taken. On the contrary, technical progress should be viewed as endogenous, as something that depends on the incentives and on the action taken in the economy" (Mäler, 1974, p. 96). These incentives can, for instance, come from Pigouvian taxes, correcting for the distortions of market failures (pollution externalities).

The recent literature on *endogenous growth theory* addresses this question. Some of the insights of this 'new' growth theory applied to environmental resource prob-lems will be summarised here[85].

---

[85]The literature on endogenous growth theory, as developed in the last five years, can be seen as a reaction to the inadequacy of the old growth models with exogenous technical progress (Ramsey, 1928; Harrod, 1948; Solow, 1956) to explain stylized facts on economic development.

The introduction of vintage models (e.g. Johanson, 1959), where technical progress is embodied in capital investment is an early improvement, though technical progress is still left exogenous. Arrow's 1962 paper on learning-by-doing is another attempt to incorporate knowledge in the growth literature. In this model, the productivity of a given firm is assumed to be an increasing function of cumulative aggregate investment for the industry. Hence, there are increasing returns

Basically, it is possible to distinguish between models with the following two types of technological progress[86]:

- technological progress that increases output without increasing energy inputs and/or without growing flows of pollution (**neutral technical change**);

- technological progress that increases the efficiency of energy inputs and/or abatement technology (**energy augmenting technical change**)[87].

A first model in the environmental literature using endogenous growth is a Greenhouse Effect model by Gottinger (1991)[88]. This model has basically the same structure as Musu (1990)[89]. The difference is that non-consumed production can be invested in either capital accumulation, K, or in 'knowledge', H. The fraction θ of investment that is spent on knowledge is a control variable. Capital and knowledge differ in two major ways, First, capital depreciates at the rate δ, while knowledge does not depreciate[90]. Second, the change in capital is a linear function of capital investment, but the change in knowledge is a non-linear function of knowledge investment. (Gottinger, op. cit. p.2). The equations of motion are thus:

$$\dot{P} \;=\; F - \alpha P \tag{2.43}$$

to scale on an industry level and decreasing returns on a firm level. Due to the public good aspect of knowledge, Arrow is able to sustain a competitive market outcome in this model with increasing returns to scale. A drawback is that the technological progress, though endogenised, is still quite mechanic in that there is a fixed proportion between 'learning' and 'doing'. Also, knowledge is assumed to be a totally unintentional side effect of the production of the conventional good.

The recent endogenous growth theory starts with the idea that technological progress is the result of deliberate decisions by the private sector to invest in knowledge, research, development, etc. (i.e. human capital). See for instance Romer (1986, 1990), Lucas (1988), Van der Klundert (1990), Barro & Martin (1990), Bean (1990) and King & Rebelo (1990).

[86]See also Gottinger (1990).

[87]See also Xepapadeas (1991, p.3).

[88]Recently, some other models on endogenous growth have been analysed. Gradus & Smulders (1993) look at different types of endogenous growth models (among others Lucas, 1988) and describe how the results change once pollution is incorporated. As pollution is considered as a flow variable only, this model will not be analysed here further. Another model on sustainable growth and the environment is by van der Ploeg & Ligthart (1993). As the focus is on strategic interactions in a two country setting, this model will be analysed in the next section where the international dimension of environmental resources is discussed.

[89]Gottinger (1991) has consumption as the only variable in the social welfare function, whereas Musu (1990) includes also the stock of pollution. Hence the only way in which pollution affects future consumption in Gottinger's model in through the production function.

[90]Gottinger (1991, p. 2).

$$\dot{H} = (\theta I)^{\sigma} \qquad\qquad 0 < \sigma \le 0 \qquad\qquad (2.44)$$
$$\dot{K} = (1-\theta)I - \delta K \qquad\qquad (2.45)$$

The production function consists of four factors: capital (K), knowledge (H), the flow of pollution (this is fossil energy, F) and a stock of pollution (P) with the following functional form:

$$Y = h(F, P)H^{\mu}K^{\nu}$$

Hence, this model is characterised by neutral technical progress in knowledge and by capital accumulation.

A recent model that looks specifically at progress in the knowledge of pollution saving techniques is by Bovenberg & Smulders (1993). This model is rather more advanced than Gottinger (1990) in many respects.

First, pollution[91] changes over time according to a concave regeneration function, $G(.)$ below, as is common in fishery models (cf. Barbier & Markandya, 1989)[92].

Secondly, Knowledge of pollution saving techniques (H) increases due to physical capital used for the 'production' of this knowledge (K_H) and energy (or harvest) used (F_H). Production can be consumed, invested or used for net capital accumulation. Output itself is produced with the aid of physical capital (K_Y) and harvest/emissions (F_Y). At the same time, output is adversely affected by the stock of pollution. Hence we have:

$$\dot{P} = G(P, F) \qquad\qquad (2.46)$$
$$\dot{H} = H(K_H, F_H) - \delta_H H \qquad\qquad (2.47)$$
$$\dot{K} = Y - C - \delta_K K \qquad\qquad (2.48)$$
$$Y = Y(P, K_Y, F_Y) \qquad\qquad (2.49)$$
$$K = K_Y + K_H \qquad\qquad (2.50)$$
$$F = F_H + F_Y \qquad\qquad (2.51)$$

Bovenberg & Smulders (1993) go on to derive the particular conditions on technology and preferences required for feasibility and optimality of balanced growth for this two-sector economy.

---

[91] Bovenberg & Smulders (1993) focus on the natural environment instead of the stock of pollution. They refer to the annual extraction out of the environment as pollution.

[92] The difference is that Barbier & Markandya assume that pollution rises over time due to consumption ('polluting output' model). In Bovenberg & Smulders (1993), it is the harvest (or energy, etc.) that affects the natural environment ('polluting input' model).

A quite different model of endogenous growth is by Hung, Chang & Blackburn (1992). The model is based on Romer (1990) with the three sectors of production: a final goods sector in which a single consumption good is manufactured; a producer goods sector in which a range of intermediate inputs are produced; and an R & D sector in which designs for new intermediate goods are created. Two types of producer goods are distinguished: an environmentally friendly good and an environmentally unfriendly good. This is an interesting first step towards modelling an economy where producers can react to shifts of consumers' preferences towards environmentally friendly goods. The main shortcoming of the model is that pollution is modeled as a flow rather than as a stock.

It is too soon to draw any general conclusions from these 'new' growth models incorporating environmental degradation. The main question is whether sustained growth with tolerable pollution is possible under plausible conditions. Both Bovenberg & Smulders (1993) and Hang, Chang & Blackburn (1992) have made a good start in trying to answer this question.

2.4 The International Dimension

2.4.1 Introduction

Up to now, we have ignored strategic elements. They may, though, be of crucial importance for the feasibility of environmental policy in practice. Situations where strategic elements can play a role are for instance the interaction between the government and the private sector and the interaction between two sovereign governments. The former have been extensively dealt with in the context of macroeconomic policy discussions (e.g. Persson & Tabellini 1991). Examples of such a form of interaction within the pollution control context are Wirl (1991), Gradus & Kort (1992)[93] and Yeung & Cheung (1992)[94].

Interactions between different governments have also been considered comprehensively in international macroeconomic policy coordination (e.g. Miller & Salmon, 1985). Following this literature, environmental economists have started to use the

---

[93]Both Wirl (1991, Appendix) and Gradus & Kort (1992) analyse a Stackelberg differential game with the government as leader. The private sector is confronted with two types of taxation, a profit tax and a pollution tax.

[94]They analyse a differential game of capital accumulation and pollution control between the government and a producer. The firm maximises output and the government's objective is to tax industrial output in order to spend the proceeds on pollution abatement.

same tools over the last few years. This has allowed the formal analysis of strategic aspects of international environmental policy. Strategic elements in pollution control arise especially in situations of renewable and environmental resources, where an externality is present due to the transboundary character of the resource (e.g. the 'global commons')[95].

In this section, as a prelude to the analysis of Chapter 4 and 5, the focus is on the strategic interaction between sovereign governments where these governments themselves are not faced with strategic conflicts domestically i.e. vis-à-vis industry and/or consumers[96].

2.4.2 International Pollution Games

The optimal control papers in this section deal especially with the problems that arise once the transboundary character of pollution[97] is taken into account. These are[98]: Hoel (1990), Cesar (1990, 1993-a), Mäler (1991), Xepapadeas (1991), Van der Ploeg & de Zeeuw (1992), Tolwinski (1992) and Van der Ploeg & Ligthart (1993).

The crucial extension compared to the 'one player' models discussed in Section 2.3 is that externalities created by spill-over effects of pollution arise in the decision making process of the social planners in the different countries. More precisely, unless cooperation on a international basis takes place, these externalities will not be internalised and a globally optimal pollution abatement strategy cannot be hoped for. The papers just mentioned capture this problem by using a Nash-framework in a differential game setting. Typically, the open-loop, feedback and Pareto solutions are compared.

The transboundary aspect means that the stock of pollution is affected by emissions in period t of all countries $(j = 1..N)$ involved: $E_t = \sum_{j=1}^{N} E_{j,t}$. These emissions are linearly related to output $(Y_{i,t})$. In the absence of cooperation, the central

---

[95]Other elements, like the link between strategic trade considerations and environmental policy, though important, will not be considered here (see D. Ulph, 1992). Again another issue is the relation between the choice of environmental policy instruments and strategic environmental trade (see A. Ulph ,1992)

[96]This does not mean that strategic interactions between government and private sector are less important. Rather, in order to concentrate on one issue at a time, they are assumed not to exist.

[97]For strategic considerations in exhaustible resources, see Reinganum & Stokey (1985) and in renewable resources (fishery), see Lehvari & Mirman (1982).

[98]Papers on international pollution games in a static game-theory setting are, among others, Mäler (1989) and Barrett (1990).

planner in country i is assumed to maximise social welfare in his country, $U_i(C_i, P)$, subject to:

$$C_{i,t} = Y_{i,t} \tag{2.52}$$

$$E_{i,t} = \epsilon Y_{i,t} \tag{2.53}$$

$$\dot{P}_t = \sum_{j=1}^{N} E_{j,t} - \alpha P_t \tag{2.54}$$

Note that for pollution problems with asymmetric dispersion, such as acid rain, a transportation coefficient has to be added to relate deposition in country i with the emissions in country j[99].

The international build-up of the stock of pollution given in Equation 2.54 is modelled in a similar fashion by most authors. The papers differ in the way the rest of the economic process in modeled. This will be briefly discussed.

The models of Hoel (1990)[100], Mäler (1991) and Van der Ploeg & de Zeeuw (1992), have basically the same structure as Dasgupta (1982) (cf. Equation 2.13). The difference is that Dasgupta (1982) additionally allows for abatement possibilities and ways to affect the assimilative capacity of nature[101].

Cesar (1990; 1993-a) and Tolwinski (1992) use the same framework but allow for abatement possibilities[102,103]. Xepapadeas (1991) and Van der Ploeg & Ligthart (1993) introduce technological progress. In Xepapadeas (1991), fossil fuel saving technological progress is considered, whereas Van der Ploeg & Ligthart (1993) analyse neutral technological progress. Investment has in both cases a public good

---

[99]If emissions are spread out equally over the globe, such as in the case of the Greenhouse Effect, Equation 2.54 is appropriate. In the case of acid rain, the corresponding equation of motion of the stock of pollution in region i is (see Mäler (1989):

$$\dot{P}_{i,t} = \sum_{j=1}^{N} a_{ij} E_{j,t} - \alpha P_{i,t}$$

where a_{ij} denotes the fraction of emission from region i that is deposited in region i

[100]Hoel (1990) assumes that the assimilative capacity of nature is zero. Herewith he reduces the problem to an exhaustible resource problem.

[101]Hence, $I = 0$ and $J(G_t, P_t) = \alpha P$.

[102]Tolwinski (1992) uses a decentralised framework in which producers maximise net profit depending on the value of output and on the total costs of emissions. The value of output depends on the production capacity, on abatement capital and on the level of pollution.

[103]Cesar (1990; 1993-a) uses a specification with investment that affects the emission-output ratio. In the first paper, both symmetric and asymmetric dispersion models are analysed.

character, which means that, at the international level, there are two externalities at the same time, one pollution spill-over and one technology spill-over[104].

The general conclusion from these models is that cooperation pays and leads to a better environment[105]. Whether output is higher or lower in the case of cooperation depends on the specification of the model.

2.4.3   International Environmental Cooperation

The aim of international negotiations on transboundary pollution problems is undoubtedly to reach an 'International Environmental Agreement' (IEA). Such an IEA can be viewed as an outcome of a game in which the following questions are solved:

- how much should be abated globally?

- how much should each country contribute to this abatement?

- how will countries be compensated for these abatement activities (cost-sharing)?

- how will free riding be deterred?

- how can compliance with the agreement be monitored?

The aim here is not to analyse these points in general. This has been done by many authors, among others Grubb (1989), Nitze (1990), Barrett (1990, 1991), Bohm (1990, 1991), Mäler (1991), Young (1991) and Parson & Zeckhauser (1992).

In what follows, two specific points will be considered: the deterrence of free riding and the idea of second-best alternatives to cost-sharing.

In the environmental economics literature, as in the case of other market failures, it is assumed that joint action requires the presence of intervention by the government. In the case of transboundary pollution, this would require a world-government. In the absence of such an identity it seems very difficult to avoid *free-riding*.

---

[104]Besides, Van der Ploeg & Ligthart (1993) consider an externality in government investments.

[105]Van der Ploeg & Ligthart (1993) show that in the case of two externalities, cooperation might adversely affect the environment. The reason is that in the absence of cooperation, there is assumed to be no international patent market and hence technological progress is impeded. In the cooperative case, patents are introduced, thereby boosting growth and hence, environmental degradation. This effect may outweigh the positive effect of cooperation on the environment.

The only way to avoid free riding is if the agreement between parties is *self-enforcing*. This means that it should be in the interest of each member of the agreement not to free ride (Barrett, Ch.3 OECD, 1991). As mentioned in Chapter 1, this can be achieved in a game theoretic framework using trigger strategies (J. Friedman, 1971)[106] and renegotiation proof strategies (Farrell & Maskin, 1989; Van Damme, 1989).

The problem of avoiding free riding in international environmental agreements has also been considered in a quite different context, by looking at coalitions of cooperating countries. A stable coalition is one in which there is no incentive to defect and no incentive to broaden the coalition (Barrett, 1989). Carraro & Siniscalco (1991) develop the same argument with the additional behavioural assumption that a subset of agents are committed to cooperation. This means that they will not defect from cooperation as long as their welfare is higher than in the Nash equilibrium. Distributing some of the surplus to others, the committed members are able to enlarge the coalition.

Parson & Zeckhauser (1992) adopt a similar approach. They take nations consisting of different-sized collections of identical individuals. The behavioural assumption is that no nation will join unless all larger nations have joined already. This creates a type of linkage. It is further assumed that those who join make equal proportional reductions from their emissions in the original Nash equilibrium[107].

Another paper discussing the issue of stability is Heal (1991). He shows that if there are fixed costs associated with abatement programs, coalitions in excess of a minimum critical size are not vulnerable to 'free riding'.

A quite different approach is taken by Chandler & Tulkens (1991). They suggest a dynamic process with a cost sharing rule. The solution of this process converges to a Pareto optimum. The idea is that the gains of cooperation are, at every step in the process, divided among the players and Chandler & Tulkens prove that this cost sharing rule has 'locally strategic stability'.

Cost-sharing is often necessary in order to obtain cooperation if there are net-losers in a coalition[108]. Such countries would never be willing to join an agreement

---

[106]The concept of trigger strategies has been used in a dynamic game framework by, among others, Tolwinski, Haurie & Leitmann (1986) and applied to natural resource management by, among others, Kaitala & Pohjola (1990).

[107]The authors prove in a ten-country world that a seven-country coalition is stable.

[108]In the case of the Greenhouse Effect, this may, for instance, be a country like Canada or Russia, that could actually benefit from some Global Warming.

where they have to incur expenses for emission reduction and if they are not compensated for doing so. Likewise, Less Developed Countries might 'lose' from signing an International Environmental Agreement that obliges them to curb emissions substantially. From an economic point of view, cost sharing is a means to ensure first best welfare results with highly asymmetric players. In this way, the question of who pays for the global emission reduction agreement becomes independent of the issue of where the control measures have to take place[109].

Due to the problems involved in finding a first best solution for global emission reduction, it may be worthwhile to look at other possibilities, that have a *second-best* character. One example is the donation of emission-reduction devices in the form of technology transfers, as an in-kind alternative to politically infeasible cost-sharing. Another example is the linkage of two unrelated, roughly-offsetting issues as a way to achieve cooperation when agreement on each individual issue, treated separately, is impossible. These two examples will be worked out in Chapter 5.

---

[109]In order to solve this redistribution problem, Hoel (1990) proposed a scheme in which the global carbon taxation revenues are reimbursed so that every country is made better off, thereby tacitly assuming that there exist 'fairness' principles bringing about this result (see also Barrett (1989)). In the case of acid rain, estimates have been carried out to see which kind of possible 'fairness' ideas lead to reimbursements so as to make everybody better off. See, for instance Mäler (1989) and Klaassen & Janssen (1989). The latter authors conclude that neither 'polluter pays', nor population nor 'equally shared responsibility' lead to solutions in which everyone is better off. Typically, only a combination of 'fairness' principles could lead to a solution that makes it profitable for each country to cooperate in such an international carbon tax scheme (see also Hoel (1991)). An interesting recent example for the case of the Greenhouse Effect is by Fujii (1990).

Another type of scheme is developed and analysed for the case of acid rain by Bergman, Cesar & Klaassen (1990). In this scheme, the 'European Environmental Protection Fund' fully pays the abatement efforts of the countries; each country contributes to the fund according to the 'minimum costs of unilaterally attaining their target level of pollution'. Note that this makes sense in the case of acid rain where different regions can have different targets and where costs can often be attained unilaterally, but that this is not feasible in global pollution problems like the Greenhouse Effect and ozone depletion.

Instead of tax schemes with reimbursements, it is also possible to obtain globally efficient pollution control via tradeable emission permits (see Tietenberg, 1985; Grubb & Sebenius, 1991). In this case the initial distribution of the permits form a challenge similar to the reimbursement scheme, where likewise countries will not cooperate if this distribution of emission permits is not in their interest.

2.5   Summary and Conclusion

In this chapter, a review has been given of the literature on environmental economics, relevant for the analysis later on in the disseration. First, the Greenhouse Effect itself was analysed, with emphasis on causes, effects as well as on greenhouse policy options.

Next, optimal control models incorporating environmental resources were surveyed. On the ecological side, different ways to model sustainability were discussed and on the economic side, the trade-off between current consumption and environmental quality was highlighted. This trade-off played a central role both when capital (physical and/or human) was assumed fixed and when it was allowed to evolve over time. Abatement, recycling, process-integrated changes as well as direct drops of production were distinguished as ways to reduce emissions. Pollutants were modeled as intermediate goods ('polluting input models') or as by-products of production ('polluting output models').

The equilibrium solution of such models is typically a unique saddlepoint with positive levels of pollution and emission reduction, although this was shown to depend crucially on the assumptions made.

In an international setting, pollution spill-overs cause equilibria to be Pareto-inefficient in the absence of cooperation. Stability issues and free rider problems, discussed in the literature, have been summarised. It appeared that the prime focus was on discussing the welfare gains of coopearation and little attention has been paid till now to the negotiation question of how to get from non-cooperative equilibria towards cooperative outcomes.

Chapter 3

Sustainability, Emission Reduction Technology and Human Capital

Fitti nel limo dicon: "Tristi fummo
ne l'aere dolce che dal sol s'allegra,
portando dentro accidïoso fummo:

or ci attristiam ne la belletta negra."[1]

Dante Alighieri (1309).

3.1 Introduction

In the previous chapter, existing optimisation models on environmental resources were reviewed. Some models looked at a single decision maker (one country) while others focused on strategic aspects arising from the introduction of international environmental resources with accompanying externalities. In this chapter, one-country models of the Greenhouse Effect are developed. Subsequently, in Chapter 4, the focus is on the transboundary aspects.

This chapter starts, in Section 2, with an elaboration of the one-state variable model of Nordhaus (1982). Specific functional forms are taken in order to analyse

---

[1]La Divina Commedia, Inferno, Canto VII, 121-124; English translation (D.L. Sayers): *Bogged there they say: "Sullen were we – we took \\ No joy of the pleasant air, no joy of the good \\ Sun; our hearts smouldered with a sulky smoke; \\ Sullen we lie here now in the black mud."*

the features of this model. The build-up of Greenhouse gases is assumed to have negative impacts on productive capacity and not on the people's well-being directly. This is a deliberate choice away from the luxury character of environmental quality. This specification implies that the fundamental intertemporal trade-off is between current consumption and future consumption rather than between current consumption and future amenity.

For a linear specification of the natural assimilation function this means that with serious enough environmental problems[2], a higher time preference leads to lower steady state consumption. This is in contrast with the luxury type models where less patience in society leads to higher steady state consumption.

Next, in order to discuss the concept of sustainability, non-linear specifications of the natural assimilation function are introduced. There exists large uncertainty with respect to the actual assimilative capacity of nature for elevated levels of the concentration of Greenhouse gases. Given this uncertainty, the robustness of policy conclusions for changes in the specification of the natural regeneration function is analysed. The conclusion is that slight variations in the parameters of the assimilation function can have a dramatic impact on the steady state results. This indicates that explicit modelling of uncertainty in future research is extremely important.

In Section 3, physical capital accumulation in the energy sector will be introduced. This allows for an analysis of the optimal level of investment in energy-related capital versus current consumption and fossil fuel use. Conditions under which the steady state is locally asymptotically stable are analysed. Comparative statics results for changes in time preference are presented. For the numerical specification of the model, more patience means in the long run a higher level of consumption, abatement capital and environmental quality.

The trajectories of the state and control variables show a considerable overshooting of energy-related capital. This means that it is optimal for an economy to start with a rapid build-up of energy technology that depreciates later on.

Finally, for the decentralised version of the model, the consequences of setting carbon taxes too low are shown to have a dramatic impact on the steady state.

In Section 4, both the human and physical capital aspects of energy-related technology are modeled explicitly. In a general specification of the models, the conditions are analysed for locally asymptotic stability of the restricted dynamic

---

[2]This holds under some conditions, analysed in the next section.

system where the level of pollution is set to its steady state level. Trajectories for human and physical capital are discussed for different starting values of these variables. The results are in line with standard two-sector growth models, where disinvestment takes place in the sector with a relatively abundant initial capital stock.

Explicit modelling of human capital is particularly important if externalities in its accumulation are present. The consequences of such knowledge spill-overs are discussed. The externalities are modeled such that in the absence of government intervention, production has decreasing returns to scale. With internalisation, there are constant returns to scale. This is an important first step to modelling endogenous growth in an economy with pollution, where sustained growth depends on economic policy. In the decentralised version of the model, the externalities in both pollution and knowledge are internalised through carbon taxes and human capital subsidies respectively.

3.2 The Rudimentary Model

3.2.1 General Description

The model of Nordhaus (1982) is the most elementary one that captures the intertemporal trade-off between consumption and Greenhouse gas build-up. The economy is assumed to have fossil fuel as a sole productive input. The use of energy in the form of fossil fuel is, however, not only a 'good' but also a 'bad' in the sense that it enhances the Greenhouse Effect. It is assumed for the time being that there are no 'abatement' possibilities[3], unlike in Forster (1977) and many others. This means that the only way to reduce the Greenhouse Effect is to diminish fuel use and hence to decrease production[4]. The model of Nordhaus (1982) will be analysed in some detail because it serves as a benchmark case for comparison with extensions of the model in later sections.

As elaborated in the last chapter, pollution can affect social welfare via a 'productivity' effect and via an 'environmental amenity' effect. Basically, the pro-

---

[3]Abatement is written in apostrophes as there are very few real abatement possibilities in the case of the Greenhouse Effect. The main ways to reduce the build-up of GHGs is to economise on fuel or to introduce non-fossil energy sources. However, the term 'abatement' seems to be used extensively in the optimal control literature on the Greenhouse Effect.

[4]In the next section, investment in energy-related technology will be introduced, which allows for a richer choice.

ductivity effect of pollution refers to, for instance, health risks to workers. The environmental amenity effect of pollution describes the case where pollution has a direct effect on social welfare.

Generally it would, in my opinion, be appropriate to analyse both amenity and productive effects. However, Greenhouse gases (esp. carbon dioxide, methane and nitrous oxides) are benign[5] and this is in contrast with most polluting gases that lead to nuisance, illness, retardation, etc. At the same time, Greenhouse gases affect the global climate in the long run, possibly leading to economy-wide disruptions (see Section 2.2). Therefore a deliberate choice is made to assume that the only effect of the build-up of Greenhouse gases is on productive capacity. Later on, this choice is shown to be of key importance for the effect of a higher time preference for future consumption.

Finally, fossil fuels are assumed to be non-exhaustible. This is obviously not true. The rationale for using this simplification is the idea that environmental constraints will come much earlier in time than the physical fuel limits (e.g. there may be enough coal for more than thousand years). A model that incorporates both the environmental and resource constraint is analysed in Forster (1980).

These considerations can be formalised as follows:

The social welfare function depends solely on consumption $U(C)$ and is assumed to be increasing and strictly concave in C (that is $U'(C) > 0$ and $U''(C) < 0$)).

Production, Y, is a function of energy use, F, and the build-up of the Greenhouse Effect, P (P of 'pollution'): $Y = Y(F, P)$. Assume that $Y(F, P)$ is a strictly concave function of F and P, and is increasing in F and decreasing in P: $Y_F > 0$, $Y_P < 0$, $Y_{FF} < 0$, $Y_{PP} < 0$ and $Y_{FF}Y_{PP} - Y_{PF}^2 \geq 0$. Besides, it is assumed that $Y_{PF} < 0$[6]

As outlined above, there are no abatement possibilities nor ways to invest in energy efficiency. Hence, the only way to reduce the concentration of CO_2 and other GHGs is to use less fossil fuels by producing less. Therefore consumption equals production: $C = Y(F, P)$.

---

[5]Most models of the Greenhouse Effect specify the feedback of pollution on future welfare through amenity effects. This typifies the situation in which environmental quality is seen as a luxury good that does not influence consumption as such though it may affect the well-being of the individual in society. Note that CFCs (by destroying the ozone layer) can have very malign effects (e.g. skin cancer)

[6]The reasoning behind this assumption is that $Y(F, P)$ will be taken to be separable into $e(F).f(P)$, where $e(F)$ is the energy function and $f(P)$ is a Greenhouse damage function, with $e' > 0$, $f' < 0$ and e and f are both concave.

The Greenhouse gas concentration is assumed to evolve over time according to the following process:

$$\dot{P} = \gamma F - \alpha P \tag{3.1}$$

This means that P increases over time due to fossil fuel use, F, with the coefficient γ denoting the percentage of fossil fuel that will end up in the atmosphere. The stock of GHGs will gradually decay due to the assimilative capacity of nature at rate α. This corresponds to assuming that the average atmospheric lifetime of GHGs is $1/\alpha$[7]. For the time being, a linear decay function αP is assumed, although in a later section on sustainability, more realistic non-linear forms will be examined. Note that reforestation is not analysed here explicitly[8].

Assume the existence of a Central Planner, who seeks to maximise the discounted flow of social welfare, which depends solely on consumption[9,10]. Hence:

$$\max_{F} \int_0^\infty e^{-\rho t}[U(C)]dt \qquad \rho > 0 \tag{3.2}$$

s.t.

$$C = Y(F, P) \tag{3.3}$$

$$\dot{P} = \gamma F - \alpha P \tag{3.4}$$

The model, as it stands here, is a so called 'fixed infinite time free right hand endpoint optimal control model'. The necessary conditions for a solution to the above model are found using Pontryagin's Maximum Principle[11]:

---

[7]Another way of saying this is that the half-life of GHGs is $1/\alpha$.

[8]This could be added in the model in the following way by assuming reforestation R with corresponding costs $C(R)$: Consumption would then be equal to: $C = Y(F, P) - C(R)$ and the evolution over time of the stock of Greenhouse gases: $\dot{P} = \gamma F - \zeta R - \alpha P$ where ζ denotes the buffering capacity of reforestation R. Note, as said before, that only continuing reforestation leads to permanent increases in the buffer-capacity.

[9]This model will be decentralised later on, in order to show that with appropriate Pigouvian taxation, a market economy can reach the social optimum.

[10]In a more recent model, Nordhaus models two state variables: the Greenhouse gas concentration (P) as well as temperature (T), where the latter is assumed to influence the economy. (Nordhaus, 1989). Tahvonen, von Storch & Xu (1992) take the same approach where, instead, $\dot{T}$ influences the economy (via the utility function). This specification makes it possible to incorporate the lag that seems to exist between emissions and actual temperature increase. This increase is ultimately the important factor for the economy. The specification with $\dot{T}$ is justified on the grounds that it might, in the end, be the speed of temperature changes that determines how badly ecosystems are affected.

[11]Note that a deterministic setting is chosen throughout. This means that the open-loop solutions are optimal. In the case of uncertainty, feedback solutions should have been considered to allow for adaptive policies.

There exist a co-state function ψ such that with the Hamiltonian defined as:[12]:

$$\mathcal{H}(F, P, \psi) = U[Y(F, P)] + \psi(\gamma F - \alpha P) \qquad (3.5)$$

the necessary conditions are given by Equation 3.4 and:

$$\frac{\partial \mathcal{H}}{\partial F} = U'Y_F + \gamma\psi = 0 \qquad (3.6)$$

$$\frac{\partial \mathcal{H}}{\partial P} = U'Y_P - \alpha\psi \qquad \text{or} \qquad \dot{\psi} = (\rho + \alpha)\psi - U'Y_P \qquad (3.7)$$

Note that, given the concavity assumptions on $Y(F, P)$ and on $U(C)$, the necessary conditions for optimality are also sufficient[13].

Equations 3.6 – 3.7 can be rewritten as:

$$U'Y_F = -\gamma \int_t^\infty U'[C(s)]Y_P[F(s), P(s)]e^{-(\rho+\alpha)(s-t)}\, ds \qquad (3.9)$$

This equation shows that at every point in time, the extra consumption due to an additional unit of fuel use equals the present value of the loss in output (in utility terms) due to the enhanced concentration of GHGs caused by this additional unit of fuel.

Assume limit conditions to hold for $U(C)$ and $Y(F, P)$, which prevent boundary solutions to occur[14]. These conditions are:

$$\lim_{C \to 0} U'(C) = \infty \qquad (3.10)$$

$$\lim_{F \to 0} Y_F'(F, P) = \infty \qquad (\forall P \text{ as long as } Y(F, P) > 0) \qquad (3.11)$$

$$\lim_{P \to 0} Y_P'(F, P) = 0 \qquad (3.12)$$

---

[12]It is also assumed that the transversality conditions are satisfied. In an infinite time horizon optimal control problem, this means that the present value of the costate approaches zero as time goes to infinity.

[13]With reforestation as modeled in the footnote above, the additional first order condition would be:

$$\frac{\partial \mathcal{H}}{\partial R} = -U'C' - \zeta\psi = 0 \qquad (3.8)$$

For the internal solution of the model, this would mean that marginal expenditures on reforestation and resulting fossil fuel reductions should be the same in the optimum.

[14]Note that limit conditions are not the same as transversality conditions, discussed in the footnote above. Besides, the limit conditions are by no means necessary for optimality. They are assumed to hold here in order to concentrate on internal solutions (see Forster (1977) for an analysis of both boundary and internal solutions).

These assumptions allow us to focus on interior solutions (with $P > 0$, $F > 0$, and $C > 0$) of the model. From the first order conditions given in Equation 3.6 the derivative of $\mathcal{H}_F$ with respect to F, P and ψ can be obtained:

$$\frac{\partial \mathcal{H}_F}{\partial F} = U'Y_{FF} + U''Y_F^2 \tag{3.13}$$

$$\frac{\partial \mathcal{H}_F}{\partial P} = U'Y_{FP} + U''Y_P Y_F \tag{3.14}$$

$$\frac{\partial \mathcal{H}_F}{\partial \psi} = \gamma \tag{3.15}$$

It is clear from the assumptions above, that $\frac{\partial \mathcal{H}_F}{\partial F} < 0$ and $\frac{\partial \mathcal{H}_F}{\partial \psi} > 0$. However, the sign of $\frac{\partial \mathcal{H}_F}{\partial P}$ is ambiguous. Here, it is assumed[15] that $\frac{\partial \mathcal{H}_F}{\partial P} < 0$.

The first order conditions above give the derived demand function for fossil fuel as implicit function of P and ψ : $F(P, \psi)$, with:

$$F_P = -\frac{\mathcal{H}_{FP}}{\mathcal{H}_{FF}} < 0 \tag{3.16}$$

$$F_\psi = -\frac{\mathcal{H}_{F\psi}}{\mathcal{H}_{FF}} > 0 \tag{3.17}$$

The signs of the derivatives of the demand function with respect to P and ψ are as expected: both an increase in the Greenhouse gas concentration P and an increase in the valuation of the environmental damage ψ push fossil fuel use F down (note that $\psi < 0$).

Next, the dynamic system 3.4 – 3.7 will be analysed:

$$\dot{P} = \gamma F - \alpha P$$
$$\dot{\psi} = \psi(\rho + \alpha) - U'[Y(F, P)]Y_P[F, P]$$

In order to analyse the phase plane (P, ψ), the derived demand function $F(P, \psi)$ is substituted into these differential equations for the state and the costate:

$$M(P, \psi) \equiv \gamma F(P, \psi) - \alpha P = 0$$
$$N(P, \psi) \equiv \psi(\rho + \alpha) - U'[Y(F(P, \psi), P)]Y_P(F(P, \psi), P) = 0$$

---

[15]A sufficient condition for this to hold is to assume that Y(F,P) is separable in an energy function $e(F)$ and a Greenhouse damage function $f(P)$ as explained above and to further assume that $U' + C.U'' > 0$, as is trivially the case for a specification of $U(C)$ such as $U(C) = \frac{1}{1-\sigma}C^{1-\sigma}$.

Then:

$$
\begin{aligned}
M_P &= & \mathcal{H}_{\psi P} &= \gamma F_P - \alpha & &< 0 \\
M_\psi &= & \mathcal{H}_{\psi\psi} &= \gamma F_\psi & &> 0 \\
N_P &= & -\mathcal{H}_{PP} &= -U''(Y_F F_P + Y_P)Y_P - U'(Y_{PF}F_P + Y_{PP}) & &> 0 \\
N_\psi &= & \rho - \mathcal{H}_{P\psi} &= \rho + \alpha - U''Y_F F_\psi Y_P - U'Y_{PF}F_\psi & &> 0
\end{aligned}
$$

Note that $N_P > 0$, due to the assumptions on $U(.)$ and $Y(.)$ and $N_\psi > 0$ due to the assumption that $\frac{\partial \mathcal{H}_F}{\partial P} < 0$. This amounts to saying that the net effect of an increase of the GHG concentration on the marginal welfare cost of the environmental degradation is negative. All the other inequalities follow also directly from the assumption made on the functions. Hence the slopes of the two loci are:

$$
\left.\frac{d\psi}{dP}\right|_{\dot\psi=0} = -\frac{N_P}{N_\psi} < 0 \text{ (downward sloping)} \tag{3.18}
$$

$$
\left.\frac{d\psi}{dP}\right|_{\dot P=0} = -\frac{M_P}{M_\psi} > 0 \text{ (upward sloping)} \tag{3.19}
$$

The two loci, giving rise to a standard saddle structure, are depicted in Figure 3.1:

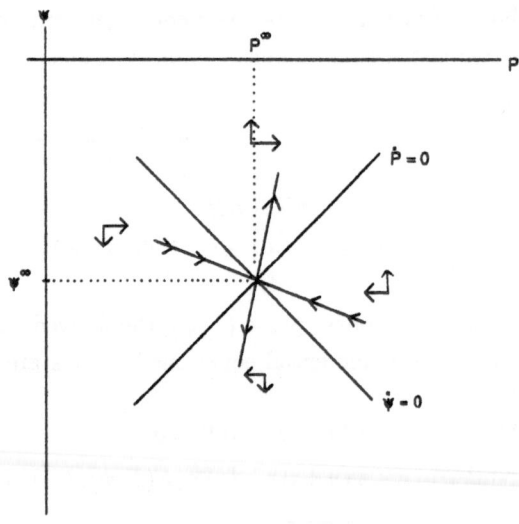

Figure 3.1: *Phase Plane of* (P, ψ)

Stability Properties The stability properties can be determined by looking at the Jacobian matrix evaluated at the steady state (P^∞, ψ^∞):

$$J = \begin{bmatrix} \rho - \mathcal{H}_{P\psi} & -\mathcal{H}_{PP} \\ \mathcal{H}_{\psi\psi} & \mathcal{H}_{\psi P} \end{bmatrix} \tag{3.20}$$

hence, det J is:

$$\det J = (\rho - \mathcal{H}_{P\psi})\mathcal{H}_{\psi P} + \mathcal{H}_{PP}\mathcal{H}_{\psi\psi} < 0$$

This means that the equilibrium is a saddlepoint and, as is clear from the slopes of the two loci, the equilibrium is unique, if it exists.

Assume in Figure 3.1, that the economy initially is left of the point P^∞. The social planner will have to set ψ in its meaning of shadow value at its corresponding level on the optimal trajectory and follow this trajectory until the steady state is reached. In the deterministic setting of the model, there is no reason why the economy would ever get off the optimal path.

Decentralisation The optimal policy of the Central Planner can be mimicked in a *decentralised* form of this rudimentary economy. Assume that there is a representative producer, a representative consumer and a government, setting taxes in the form of emission charges, τ. The producer maximises profit $\pi = Y(F, P) - \tau F$, where τ is the externality charge set by the tax collector (commodity prices are set equal to unity). Assume furthermore that the consumer maximises utility $U(C)$ (Equation 3.2 above) given his/her budget-constraint $C = \pi + T$, where the government redistributes taxes $T = \tau F$. For each value of τ, there will be an equilibrium in this economy. The social optimum in this economy will be reached if the government sets emission charges in the following way[16]:

$$\tau = -\psi \frac{\gamma}{U'} \tag{3.21}$$

This means that the optimal Pigouvian tax, τ, equals the shadow value in utility terms of environmental degradation due to the Greenhouse Effect[17].

The decentralisation shows that a government charging the optimal Pigouvian pollution tax can keep a decentralised market economy at the social optimum.

---

[16]With this level of τ, maximising behaviour of consumers and producers gives exactly the same solution as is reached in the Central Planner model (and is therefore socially optimal).

[17]Equation 3.21 can be rewritten as: $\tau U' = -\psi\gamma$. The right-hand side of this equation is the valuation of additional pollution due to an extra unit of fossil fuel. The left-hand side expresses the evaluation of the extra consumption resulting from an additional unit of fuel.

This decentralisation problem will be elaborated in greater length once capital is introduced in the model.

Note that any environmental policy resulting in the same equilibrium valuation of pollution τ gives exactly the same optimal trajectory and steady state. Hence, tradeable emission permits (with the amount of vouchers corresponding to τ), subsidies on emission reduction (at level τ) and Pigouvian taxes (at level τ) mimic the social optimum of the Central Planner when $\tau = -\psi\gamma/U'$[18].

The Steady State   In the steady state at the intersection of the two loci the two following expressions hold:

$$\gamma F^\infty = \alpha P^\infty \tag{3.22}$$

$$U'(C^\infty)Y_F = -\gamma \frac{U'(C^\infty)Y_P}{\rho + \alpha} \tag{3.23}$$

Both steady state conditions are straightforward neo-classical optimality conditions. Equation 3.22 means that, in equilibrium, as much fossil fuel is put into the atmosphere as nature can absorb. This is one fundamental idea of *sustainable* use of environmental resources, which will be discussed in more detail below.

Equation 3.23 says that the marginal welfare gains of energy use equal the discounted value of the marginal loss in welfare due to the detrimental effects of energy for the Greenhouse Effect[19]. The discount rate is $\rho + \alpha$. It may, at first sight, be surprising that α is part of the overall discount rate. Note, however, that the effects of energy use on the Greenhouse Effect are mitigated by the fact that nature absorbs a fraction α of the additional concentration of GHGs. The present value of the damage is, hence, the annual flow divided by $(\rho + \alpha)$. This highlights the crucial role of the level of the discount rate, indicating the impatience of society, for the steady state level of consumption and pollution.

This issue of time-preference can be analysed by looking at the comparative statics results for changes in ρ. Taking total derivatives of Equations 3.22 and 3.23,

---

[18]Note that for initial levels of the GHG-concentration (P_0) below the steady state (P^∞), ψ will increase in absolute terms over time. This means that r will rise as well over time (As $\tau = -\psi\gamma/U'$, emission charges τ will increase more than the shadow price ψ). This could be interesting for the discussion as to whether carbon taxes should rise or fall over time. I will, however, not go into this discussion.

[19]Equation 3.9 gives the corresponding optimality condition for any point on the trajectory.

and using Cramer's rule, the following expressions are obtained:

$$\frac{\partial F}{\partial \rho} = -\frac{\alpha Y_F}{\Delta} > 0$$
$$\frac{\partial P}{\partial \rho} = -\frac{\gamma Y_F}{\Delta} > 0$$

where $\Delta = \gamma((\rho + \alpha)Y_{FP} + \gamma Y_{PP}) + \alpha((\rho + \alpha)Y_{FF} + \gamma Y_{PF}) < 0$. The signs indicate that more impatience in society implies a higher steady state level of both fossil fuel and of environmental degradation. This is not surprising, as a higher social discount rate means a higher appreciation of current fossil fuel use (and hence production) and less concern for long run detrimental environmental repercussions. The impacts of a change in the discount rate on consumption are less clear. As $C = Y(F, P)$, it follows that:

$$\frac{\partial C}{\partial \rho} = Y_F \frac{\partial F}{\partial \rho} + Y_P \frac{\partial P}{\partial \rho} \tag{3.24}$$

As $Y_F > 0$ and $Y_P < 0$, the sign of $\frac{\partial C}{\partial \rho}$ is not known beforehand, because of the opposing effects of higher fossil fuel use and higher environmental degradation on production and hence on consumption[20].

Substituting Equations 3.2.1 – 3.2.1 in the Expression 3.24 gives the conditions for the sign of $\frac{\partial C}{\partial \rho}$ to be negative:

$$\frac{\partial C}{\partial \rho} < 0 \qquad \text{iff} \qquad \alpha Y_F + \gamma Y_P < 0 \tag{3.25}$$

This means that more patience leads to more consumption in future as long as the detrimental effects of fossil fuel use on future environmental impacts are large enough. This can be explained in an alternative way by noting that abstaining from the use of fossil fuel now is investing in a 'greener' and hence (*ceteris paribus*) more productive future. This means that $\frac{\partial C}{\partial \rho} < 0$ as long as the investment aspect is more important than the short run consumption aspect. Such situations are worked out graphically in the next section, where numerical forms of the functions are used.

3.2.2 A Numerical Specification of the Model

In the last section, a general formulation was given of a rudimentary economy affected by the Greenhouse Effect. In this section, specific functional forms are

---

[20]Note the importance of how the assimilation function is modelled. If, for instance, the natural purification process is αP for low values of P and constant $\alpha \bar{P}$ for values $P \geq \bar{P}$, then $\frac{\partial F}{\partial \rho} = 0$ in the long run if $P^\infty > \bar{P}$.

chosen to enable a more detailed analysis of some of the features of this economy. To this end, the production function is now assumed to be separable of the form:

$$Y(F, P) = e(F)f(P) \tag{3.26}$$

where e(F) is the energy function and f(P) the Greenhouse damage function. These functions as well as the specified social welfare function will be briefly described.

The Energy Function

It is assumed that there are decreasing returns to fuel in a production function in which all other elements (labour, physical capital, human capital) are fixed at the constant k. This function gives potential output in the absence of a Greenhouse problem as:

$$e(F) = kF^{\delta} \tag{3.27}$$

The Greenhouse Damage Function

The function $f(P)$ denotes the fraction of potential output that is not destroyed by the Greenhouse Effect so that the amount $e(F)(1 - f(P))$ is lost. Therefore, a specification of the damage function is chosen so that $f(P)$ is equal to 1 for $P = 0$ and declines concavely:

$$f(P) = \frac{\hat{P}^2 - P^2}{\hat{P}^2} \qquad\qquad P \le \hat{P} \tag{3.28}$$

where P is taken such that it does not exceed $\hat{P}$ (see Figure 3.2).

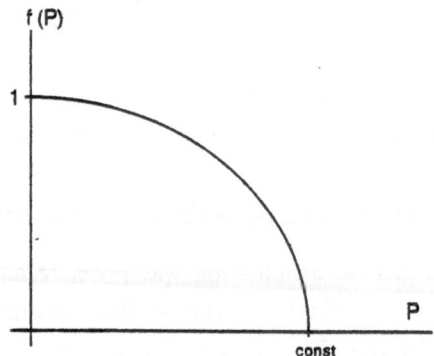

Figure 3.2: *Greenhouse Damage Function* $f(P)$

The Social Welfare Function

The social welfare function is taken to be:

$$U(C) = \frac{1}{1-\sigma}C^{1-\sigma} \qquad \text{with } 0 < \sigma < 1 \qquad (3.29)$$

This is the well-known constant relative risk aversion utility function or, more aptly since there is no uncertainty in this model, the welfare function with constant elasticity of marginal utility.

With these specifications, the Central Planner solves the following system:

$$\max_{F} \int_0^\infty e^{-\rho t} \frac{1}{1-\sigma} C^{1-\sigma} dt \qquad \rho > 0$$

s.t.

$$C = kF^\delta(\tfrac{\bar{P}^2-P^2}{\bar{P}^2})$$
$$\dot{P} = \gamma F - \alpha P$$

This model can be simply solved once again using the Pontryagin principle (see Section 3.2.1).

Numerical Values for the Parameters

In order to further analyse the model, values for the parameters α, γ, ρ, δ and σ have to be chosen.

α The parameter α indicates the percentage of the total concentration of GHGs absorbed by the environment per year. As indicated in the last chapter, this percentage is estimated at about 0.5%. Hence, α is taken to be 0.005

γ The parameter γ denotes the fraction of fossil fuel use that ends up in the atmosphere. As explained in the last chapter, scientists claim that about half of the anthropogenetic fossil fuel is dispersed in the atmosphere and the other half is absorbed by the oceans. Therefore, γ is set to 0.5. In a variation of the model, the steady state is calculated for $\gamma = 1.0$ in order to get an idea of the impact of a change of this parameter on the economy.

ρ The social discount rate (in real terms) is set to $\rho = 0.03$ in the benchmark case with $\rho = 0.01$ and $\rho = 0.06$ as variations. The choice is, of course, rather arbitrary, and as is shown later, the steady state values differ considerably with the various values of ρ.

δ The fossil fuel elasticity of production δ is assumed to lie between zero and one. As a benchmark case it is set to 0.3 and in a variation of the model, it is set to 0.6. Admittedly, this value is very arbitrary. For instance, Gottinger (1990, p.29), has set δ much higher (between 0.5 and 1.0)[21]

σ The elasticity of marginal utility is set to 0.75. Note, however, that the steady state values of C, F and P are not influenced by the choice of σ. The only variable that changes is the costate ψ.

Given the choice of the functional forms and the parameter values[22], the steady states and the trajectories of the relevant variables can be calculated. Welfare evaluations and phase-plane analysis are first elements towards an ultimate policy discussion on the basis of the model. Given the specification of the utility function as $U(C)$, welfare evaluations are straightforwardly linked to the values of consumption. The level of steady state consumption for different parameter values is discussed below and given in Table 3.1. The phase diagram is depicted in Figure 3.3[23].

As mentioned above in the section on decentralisation, ψ will increase in absolute terms over time if the initial level of the GHG-concentration (P_0) is below the steady state (P^∞). In a case with carbon taxation (τ), this means that τ will rise as well over time (As $\tau = -\psi\gamma/U'$, emission charges τ will increase more than the shadow price ψ).

---

[21]Actually, Gottinger sets the product of δ and σ to 0.50 in one version and to 0.95 in another version of his model. Given that $\sigma \leq 1$, it is obvious that δ is between 0.5 and 1.0.

[22]Apart from the parameters described above, two more parameters have to be given values: k in the production function and $\hat{P}$ in the Greenhouse damage function. The constant k is set to 17.42257. The rationale behind this choice is that with this value, the rudimentary economy gives the same outcomes of the principle variables as the model in the next section where k is replaced by a function of the new energy-related technology variable. The constant $\hat{P}$ is set to 1000. This implies that with P currently at about 350 ppm, it is assumed that the GHG-concentration will not grow beyond a tippling of current values. Note that this is, of course, a quite arbitrary number. Increasing this number with 1% leads to an equiproportional rise in P and F and hence in a $(1.01)^{0.3} - 1 \approx 0.3\%$ increase in C.

[23]Note that this figure has the same structure as Figure 3.1. The graph is made with the use of MacMath, a computer package that calculates and plots the trajectory and saddle points of two-dimensional phase-planes. The results have been compared with those obtained by the APREDIC-programme of Wymer, discussed below. The trajectories of the variables are identical for both models. The advantage of MacMath is that the input of formulae is very user-friendly. Besides, the output of MacMath is directly plotted, whereas in APREDIC, no graphics programme exists. On the other hand, APREDIC can be used for much larger systems of dynamic equations.

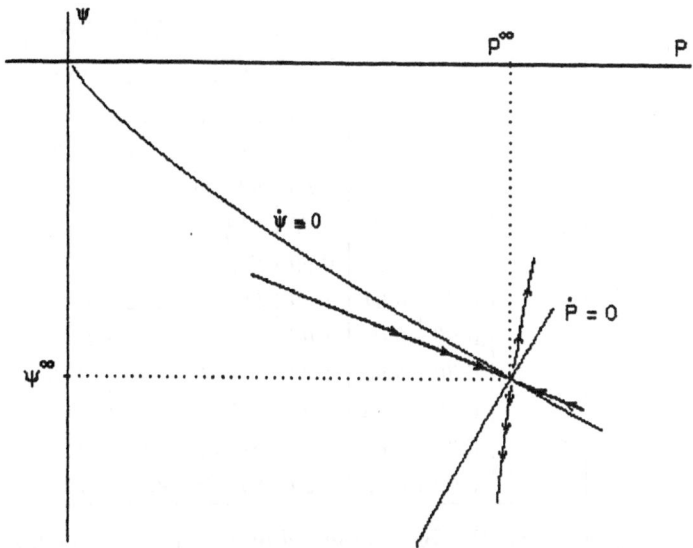

Figure 3.3: *Phase Plane of (P, ψ) for numerical values in text*

In Table 3.1, the steady state values of consumption (C), fossil fuel use (F), and environmental degradation (P) are given for the various parameter values discussed above. The benchmark case is case 3 in the table.

Note that the actual values of P^∞, C^∞ and F^∞ do not have a direct interpretation, as they are not indexed in any way. The important question is how these values change as the parameters vary and how they change over time. Comparing the benchmark case with case 4, note that a doubling of the fossil fuel elasticity of production δ increases both P^∞, C^∞ and F^∞. This might come as a surprise as a higher δ means that the same amount of consumption can be reached with less fuel use and with a lower level of pollution. On the other hand, given that fossil fuel is more productive, the economy can afford to use more. The repercussions from the increased Greenhouse gas emissions will be more than offset by the increase in potential output, thereby increasing consumption, as the results in the table indicate.

Comparing the benchmark case with case 9, where $\gamma = 1.0$, note that the steady state value of environmental degradation, P^∞, has remained unchanged. Furthermore, F^∞ has halved and C^∞ has decreased with $(1 - 0.5^{0.3})100\%$. The latter is not surprising given that $C = kF^{0.3}f(P)$. The fact that the equilibrium value P^∞ does not change follows from the specification of the model.

case	δ	ρ	γ	P^∞	F^∞	C^∞
1	0.30	0.01	0.50	557.09	5.57	20.12
2	0.60	0.01	0.50	688.25	6.88	29.17
3(bm)	0.30	0.03	0.50	715.68	7.16	15.33
4	0.60	0.03	0.50	823.05	8.23	19.91
5	0.30	0.06	0.50	813.03	8.13	11.07
6	0.60	0.06	0.50	892.14	8.92	13.22
7	0.30	0.01	1.00	557.09	2.79	16.34
8	0.60	0.01	1.00	688.85	3.44	19.25
9	0.30	0.03	1.00	715.68	3.58	12.46
10	0.60	0.03	1.00	823.05	4.12	13.13
11	0.30	0.06	1.00	813.03	4.07	9.00
12	0.60	0.06	1.00	892.03	4.46	8.72

Table 3.1: P^∞, C^∞ and F^∞ for a variety of parameter values of δ, ρ and γ; Case 3 is the benchmark case

Comparing the benchmark case (bm) with Case 1 and 5 gives the most remarkable results. The intuition behind the fact that C^∞ decreases as ρ goes up will become clear if the phase plane of the last section is drawn with locus $\dot{C} = 0$ instead of $\dot{\psi} = 0$. Figure 3.4 depicts this (P, C)-phase plane[24] for different values of ρ[25]. From the graph it is clear that the $\dot{P} = 0$–locus first increases and after a certain point decreases. This is due to the fact that $f(P)$ will go to zero for large P, thereby dominating the other variables and pushing back to the horizontal axis. It is interesting to observe that the steady state value of C is at its optimum for $\rho = 0$. The more impatient this economy becomes, the lower is C^∞ and the higher is P^∞. As was shown in the comparative statics calculations in the last section, this result comes from the importance of cutting fuel use in the present as an investment in a 'green' and productive economy in the future[26]. This can be explained by looking at the trajectory for the cases $\rho = 0.03$ (the *impatient* economy) and $\rho = 0.00$ (the infinitely *patient* economy). Assume that the economy is initially at point P_0. The impatient economy chooses a higher level of current consumption

---

[24]The differential equations of $\dot{C}$ and $\dot{P}$ can be calculated by differentiating equation 3.6 and substituting ψ and F out of the $\dot{C}$ and $\dot{P}$ equations.

[25]Note that $\rho = 0.06$ is not drawn. The steady state value for this case is for C at a point between 0 and $C^\infty_{|\rho=0.03}$. The corresponding value of P^∞ is in between $P^\infty_{|\rho=0.03}$ and $\hat{P}$.

[26]Note that $\frac{\partial C}{\partial \rho} > 0$ is not excluded in the general specification of the model.

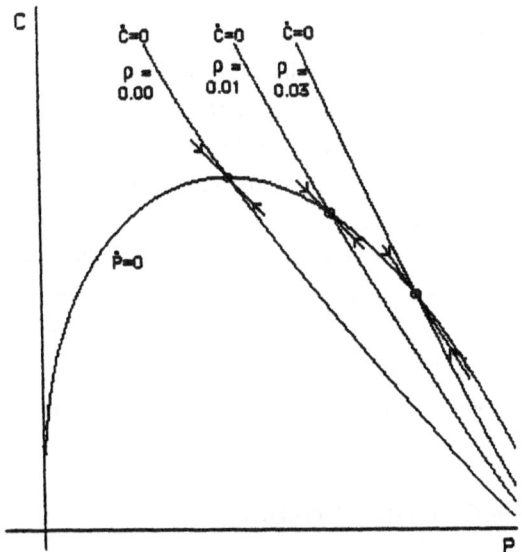

Figure 3.4: *Phase plane of* (P, C) *for different values of* ρ

(and production) at the expense of a lower steady state level of consumption (and production), whereas the patient economy sacrifices current consumption (and production) in order to have a better environment and a higher level of consumption for future generations.

This result is similar to the one in fishery economics that a higher discount rate often leads to a lower equilibrium quantity of fish and hence of income for the fishermen (see Pearce & Turner, 1990). The idea that the sustainable level of a resource stock depends on the discount rate taken highlights the importance of the 'proper' social discount rate, especially in the light of ideas about intergenerational equity (See Chapter 2).

3.2.3 Sustainability

There is a fair amount of ambiguity over the meaning of the concept of sustainability. This concept will be operationalised for different specifications of the assimilation function of Greenhouse gases. Also, the issue of survivability will be considered.

As elaborated in Section 2.3.2, the concept of sustainability is defined here as the requirement of generating pollution at rates "less than or equal to the rates at which they can be absorbed by the assimilative capacity of the environment"

(Barbier & Markandya, 1989, p.3).

For Nordhaus' (1982) model of the Greenhouse Effect, the sustainability constraint implies that the long term flow of GHGs into the atmosphere (i.e. γF) is less than or equal to the rate at which it can be absorbed by the assimilative capacity of the environment (i.e. αP). This amounts to saying that in the evolution over time of the GHG concentration, $\dot{P} = \gamma F - \alpha P$, sustainability constrains $\dot{P}$ to be less than or equal to 0 in the long run. Note that in the previous calculations of the steady states (where $\dot{P} = 0$, $\dot{\psi} = 0$), the sustainability constraint was satisfied. In these calculations, however, a very simple specification of the assimilative capacity of nature was considered. In the next subsection, attention is focused on more realistic functional forms to examine the implications of the sustainability restriction.

The Assimilative Capacity of Nature

As discussed in Chapter 2, relatively little is known about the assimilation process of the environment. Ideally, the uncertainty with respect to the natural regeneration is modelled. This is, however, rather difficult, because it is not just the maximum carrying capacity of the environment that is uncertain, but also the whole process of assimilation. Therefore, the choice is made here to model different non-linear specifications of the regeneration function in a deterministic setting and to calculate how robust the results are for variations in this specification.

As a benchmark case, the assimilation function (αP with $\alpha = 0.05$) of the previous section is taken. This numerical specification corresponds with an average atmospheric lifetime of GHGs of 200 years. This half-life holds approximately for current levels of the GHG concentration.

For levels of 'pollution' above the current level, non-linearities in the assimilation function cannot be ruled out (Arrhenius & Waltz, 1990). In order to analyse this situation, the following functional form is chosen:

$$\alpha(P) = 0.005P - aP^3 \qquad (3.30)$$

In the standard case, $a = 0$. This will be referred to as case '0'. Next, a will be set to the value 50e-10 in case '50', 51e-10 in case '51', 52e-10 in case '52' and 53e-10 in case '53'. The rationale for this specification is that it is virtually identical to the linear assimilation function for small values of P and the differences between the five only appear for higher values of P[27]. The five cases are depicted in Figure 3.5.

---

[27]Other specifications of the assimilation function are given in Cesar & de Zeeuw (1994). These

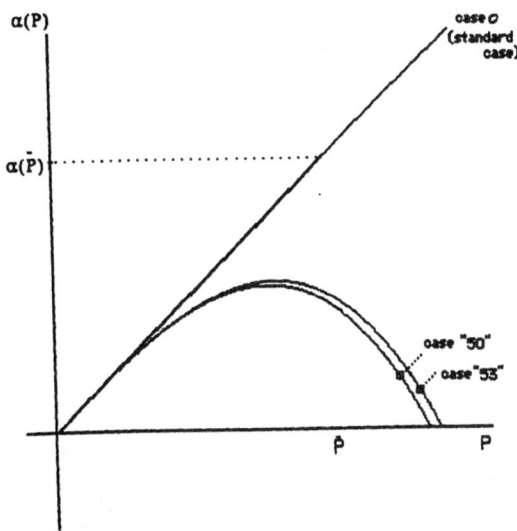

Figure 3.5: *Assimilation for different specifications of $\alpha(P)$*

Interestingly, the stability and uniqueness issues are remarkably different for the 5 cases[28]:

case '0' This is the standard scenario of the previous section with a unique and stable steady state;

case '50' This scenario has a unique steady state, which has a higher P^∞ than in the standard case;

case '51' In this case there are two equilibria, one stable and one unstable. Assuming as before that the initial value of P is small enough, the steady state is the equilibrium with the lowest level of P. However, if the initial value of P is for some reason very large, there is no way to stop a collapse of the ecosystem to take place;

case '52' This scenario has one equilibrium, that is a saddlepoint if reached from below and an unstable equilibrium if reached from above. This means that for initial values above P^∞, the ecosystem will again break down;

---

include a scenario in which the assimilation function is linear up to a certain point $\bar{P}$, after which assimilation is constant and another scenario where the ecosystem will break down above $\bar{P}$.

[28]Note that the experience with the Greenhouse effect up till now is only with low levels of P. The values of the non-linear function as such are arbitrarily chosen.

case '53' In this case, there is no intersection of the two loci. This means that
there is no saddlepoint equilibrium and the economy will eventually collapse.

The phase plane of (P, ψ) for the different cases is shown in Figure 3.6. Note that

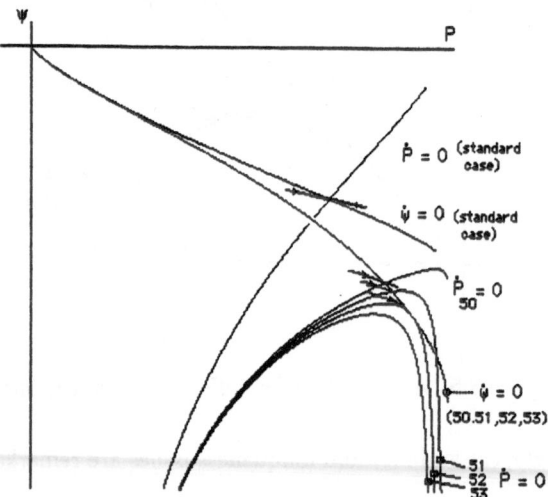

Figure 3.6: *Phase Plane of (P, ψ) for different non-linear specifications of $\alpha(P)$*

the $\dot{\psi} = 0$–locus is basically not visibly different for the four non-linear cases and
is slightly steeper than in the standard case. The difference in the $\dot{P} = 0$–locus for
these four cases is quite remarkable for high values of P, where the non-linear term
starts to dominate.

The functional form and the parameter-values chosen here are, of course, arbitrary. The sole reason why they are elaborated is to show that the assimilation
function may seem quite similar for low values of P in different scenarios but lead
to surprisingly different results when the non-linear component starts to dominate.

The main conclusion is the steady state results are not at all robust with respect
to slight variations of the linearity assumption. This also means that future research
should concentrate on trying to get a better grip on the assimilation function for
large values of pollution and on explicitly modelling the uncertainty. At the same
time, irreversibility could also be modelled explicitly[29].

---

[29]See Pethig (1990), who models the possible irreversible loss of the resource characteristic in
situations where there is prospect of better information in future.

Notwithstanding these points, the assumption of perfect foresight and of linearity of the assimilation function will be made again in the models of the next section where a capital stock is introduced. Given the fact that this last assumption was shown above to be non-robust, the policy analysis is should be done with extreme care and sound scepticism. First, however, the concept of survivability is discussed.

The Survivability of Mankind

Maximisation of intertemporal welfare using the sustainability concept, given in Section 3.2.3 could mean that the steady state level of consumption is below some threshold level $\bar{C}$. Assume that this level is the poverty line, under which the population enfeebles and eventually dies out. Then a steady state consumption below $\bar{C}$ means that the population dies out sustainably. The point is that the survivability of the population is not captured in the definition of sustainability of the Greenhouse Effect given above.

There are several ways out. Firstly, it can be argued that besides a sustainability constraint on the environmental resource (here: Greenhouse Effect), there is a sustainability constraint on population size. However, as population is not explicitly modelled here, it is imposible to use this constraint directly. Secondly, other sustainability concepts can be used that encompass the survivability of mankind (see Brown et al.; 1987)[30].

Thirdly, the social welfare function can be adjusted in such a way that a steady state consumption level below $\bar{C}$ becomes suboptimal. Here, instead, yet another way out is chosen: survivability of mankind is taken as an additional constraint in the model. In this way the sustainability concept of Barbier & Markandya (1989) is maintained but consumption below the poverty line is avoided. In Figure 3.7 an example is given where the survivability constraint is binding. This is depicted in a (P, C)-phase plane, where the assimilation function is supposed to be linear for convenience. At t_0, the concentration of P is such that the corresponding level of C_0 is above $\bar{C}$. Furthermore, assume that the steady state without the survival constraint is below the threshold $\bar{C}$.

As is clear from the figure, survivability forces the trajectory to be below the unconstrained consumption path. The result of this 'Spartan' retention is, however, that consumption levels for future generations will not be below the threshold level of survival[31].

---

[30]See Pezzey (1989) for other definitions.

[31]An interesting issue occurs when the constraints of survivability and sustainability do not

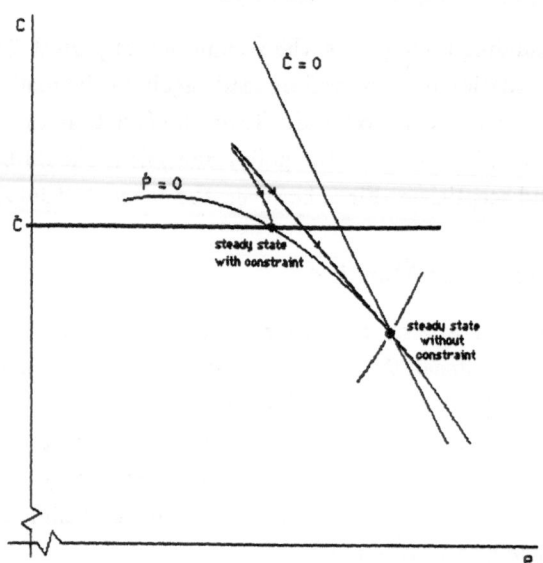

Figure 3.7: *Phase Plane of (P, C) under Survivability Constraint*

In what follows, it will be assumed that survivability is not a problem, so that the focus is on the sustainability constraint.

3.3   Energy Related Capital

3.3.1   General Description of the Model

In this section, capital accumulation is introduced in the rudimentary economy of the previous section. This allows for other means of emission reduction than by simply lowering output. In fact, abatement, recycling and process-integrated changes brought about by the introduction of capital make it possible 'de-link' economic growth and energy use.

Many authors have introduced abatement possibilities both in models with fixed capital stocks and in models that allow capital to evolve over time (cf. Chapter 2). Here, focus will be on the latter, following Keeler, Spence & Zeckhauser (1971), Brock (1977), Decker (1982), Luptáčik & Schubert (1982), Musu (1990, 1992), Pethig (1990), Tahvonen & Kuuluvainen (1990) and Van der Ploeg & Withagen (1991). These models, as elaborated in Section 2.3.4, are basically Ramsey models

---

allow for a steady state to take place. This will not be worked out here.

with pollution as an additional stock, where capital can be used for the production of both the consumption good and of abatement services[32]. The authors have mainly concentrated on steady state properties and on comparative statics analysis.

In this section, the focus is on the situation where capital is specifically a stock of energy technology[33] and/or renewable energy capital and the purpose is to economise on fossil fuel use. Note that this technology can be broadly defined, including both physical capital related to energy as well as the human capital aspect of technology.

First, the stability of the general form of the model is investigated and comparative statics analysis carried out for variations in the time preference. Then, for a numerical specification, the trajectories of the state and control variables are calculated. As in the rudimentary economy described above, changes in the time preference are analysed for the numerical form of the model. Finally, for the decentralised model, the consequences of employing levels of carbon taxes that are too low are analysed.

In order to model technological progress in the emission reduction sector, we have specified the production function as: $Y = Y\{e(F,T), P\}$ with $Y_e e_F > 0$, $Y_e e_T > 0$, $Y_P < 0$ and with $Y(.)$ being concave in F, T and P. In this specification, $e(.)$ is the energy function, as before. Now, however, energy use depends both on fossil fuel use, F, and on emission reduction technologies, T in its widest meaning, including the result of investment in energy efficiency and renewable resource development[34].

Note that we have left the traditional inputs labour and capital out of the production function in order to concentrate crucially on fossil fuel use and broadly defined energy-related technology/capital. Hence, it is implicitly assumed that investments in energy technologies do not influence the level of productive capital in

---

[32]Luptáčik & Schubert (1982) define alternatively two capital stocks, one for productive capital and one for abatement capital.

[33]By technology, we mean the actual physical capital used and not the blue-print to make it.

[34]As mentioned above, the variable F is referred to as fossil fuel use in the production function $Y(.)$. However, it is worthwhile to stress that $Y(.)$ can also be viewed as a function of joint production with F being the undesirable output (i.e. GHG emissions) (see Tahvonen & Kuuluvainen (1990)). Alternatively, following Pethig (1990), F can be seen as an input, being the demand for 'assimilation services' in the industry. In this case, $\gamma F - \alpha P$ can be viewed as the "industry's excess demand for nature's assimilative services" (Pethig, 1990, p. 5). This means that the model described here can be used for other purposes than the analysis of the Greenhouse Effect with fossil fuel and energy-related capital.

the economy. It remains to be seen whether this crucial assumption holds approximately in the real world.

As before, fossil fuel emissions lead to an accumulation in the concentration of Greenhouse gases (P), which harms economic output. The problem that the Central Planner has to solve can be formalised as[35]:

$$\max_{C,F} \int_0^\infty e^{-\rho t}[U(C)]dt \qquad\qquad \rho > 0 \qquad\qquad (3.31)$$

s.t.

$$C \;=\; Y - I \qquad\qquad (3.32)$$
$$Y \;=\; Y\{e(F,T),P\} \qquad\qquad (3.33)$$
$$\dot{P} \;=\; \gamma F - \alpha P \qquad\qquad (3.34)$$
$$\dot{T} \;=\; I - d_T T \qquad\qquad (3.35)$$

Using Pontryagin's Maximum Principle, the necessary conditions for a solution to the model are:

There exist co-state functions ψ for environmental degradation and ξ for energy technology accumulation, such that[36]:

$$\mathcal{H}(C,F,P,T,\psi,\xi) = U(C) + \psi(\gamma F - \alpha P) + \xi[Y\{e(F,T),P\} - C - d_T T]$$

With:

$$\frac{\partial \mathcal{H}}{\partial C} \;=\; U'(C) - \xi = 0 \qquad\qquad (3.36)$$

$$\frac{\partial \mathcal{H}}{\partial F} \;=\; \gamma\psi + \xi Y_F = 0 \qquad\qquad (3.37)$$

$$\frac{\partial \mathcal{H}}{\partial P} \;=\; \xi Y_P - \alpha\psi \qquad \text{or} \qquad \dot{\psi} = (\rho + \alpha)\psi - \xi Y_P \qquad\qquad (3.38)$$

$$\frac{\partial \mathcal{H}}{\partial T} \;=\; \xi(Y_T - d_T) \qquad \text{or} \qquad \dot{\xi} = (\rho + d_T - Y_T)\xi \qquad\qquad (3.39)$$

With the concavity assumptions stated, the necessary conditions are also sufficient. As in the rudimentary model, limit conditions are assumed to hold for C, F, P (cf. Equation 3.10, 3.11 and 3.12) and additionally for T. This enables the concentration on internal solution(s) of the model. Also, the transversality conditions, stating that the present value of the co-states is zero in the long run, is satisfied.

---

[35] Reforestation can easily be incorporated in this model in the same way as in Section 3.2.1.

[36] For convenience, the derivatives $Y_e e_F$ and $Y_e e_T$ will be written as Y_F resp. Y_T. Note also that I is substituted out of the system

From the first order conditions, we know that $\mathcal{H}_C = U'(C) - \xi = 0$ and $\mathcal{H}_F = \gamma\psi + \xi Y_F = 0$. These static conditions mean that a unit of value spent on either fossil fuel or on energy-related technology should, on the margin, make the same contribution to welfare.

Taking the derivatives of $\mathcal{H}_F$ and $\mathcal{H}_C$ with respect to C, F, P, T, ψ and ξ gives the derived demand functions for fossil fuel and for consumer goods as implicit function of the state and co-state equations:

$$C(\xi) \qquad \text{with} \quad C_\xi < 0$$
$$F(P,T,\psi,\xi) \quad \text{with} \quad F_P < 0 \,,\, F_T > 0 \,,\, F_\psi > 0 \text{ and } F_\xi > 0$$

Substituting these derived demand functions in the differential equations for the state and co-state equations gives:

$$
\begin{aligned}
\dot{P} &= \gamma F(P,T,\psi,\xi) - \alpha P &&= \mathcal{H}_\psi \\
\dot{T} &= Y[e\{F(P,T,\psi,\xi),T\},P] - C(\xi) - d_T T &&= \mathcal{H}_\xi \\
\dot{\psi} &= (\rho + \alpha)\psi - \xi Y_P[e\{F(P,T,\psi,\xi),T\},P] &&= \rho\psi - \mathcal{H}_P \\
\dot{\xi} &= (\rho + d_T)\xi - \xi Y_T[e\{F(P,T,\psi,\xi),T\},P] &&= \rho\xi - \mathcal{H}_T
\end{aligned}
\tag{3.40}
$$

In order to study the stability properties of this system, the Jacobian matrix of the linearised system is evaluated at the steady state $(P^\infty, T^\infty, \psi^\infty, \xi^\infty)$. It is shown in Appendix A.1 that the dynamical system 3.40 is a saddle point if the rate of time preference ρ is sufficiently low.

Interpretation of Conditions of Steady State In the steady state[37] the derivatives of the state and costate variables are set to zero ($\dot{P} = 0, \dot{T} = 0, \dot{\psi} = 0, \dot{\xi} = 0$). This means that at the steady state, the following conditions hold:

$$\gamma F^\infty = \alpha P^\infty \tag{3.41}$$

$$Y^\infty = C^\infty + d_T T^\infty \tag{3.42}$$

$$\frac{1}{\gamma} U'(C^\infty) Y_F = \frac{1}{\rho + \alpha} U'(C^\infty) Y_P \tag{3.43}$$

$$\rho + d_T = Y_T\{e(F^\infty, T^\infty), P^\infty\} \tag{3.44}$$

These steady state conditions are once again straightforward neo-classical optimality conditions[38]. Equation 3.41 is the same as in the economy without capital. It

---

[37]Though the term steady state is used throughout, concentration is on the special case with zero long-term growth (stationary state).

[38]These results are obtained by using the static first order conditions to substitute the co-state variables out of the equations. Note that substituting Equations 3.36 – 3.37 into the Equations 3.38

says that, in equilibrium, nature absorbs exactly the amount of fossil fuels coming into the atmosphere. This is the basic idea of sustainable use of environmental resources, discussed above.

Equation 3.42 says that at the steady state, there is no change in the capital stock. Hence, output can be allocated to consumption and to replacement investment alone.

Equation 3.43 is again as in the rudimentary model. It says that the marginal welfare gains of energy use equals the discounted value of the marginal loss in welfare due to the detrimental effects of energy on the Greenhouse Effect.

Equation 3.44 says that at the steady state, there is no change in the valuation of the capital stock. Hence, the marginal product of capital equals the sum of social discount rate and fraction of investment replaced.

Next, the analysis of the steady state will be used for a comparative statics analysis, with focus on changes in the social discount rate (see Van der Ploeg & Withagen, 1991; Musu, 1992).

Comparative Statics Musu (1992) gives sufficient conditions for a reduction in the social discount rate to lead to both a higher capital stock and a better environment. He uses a Ramsey model with additionally an environmental stock variable. Capital can both be used for productive purposes and for pollution abatement. Musu's sufficiency condition says basically that the increase in capital should be allocated in such a way between productive and non-productive purposes that the overall pollution goes down with increases of the capital stock (in terms of our model, this condition says that $\mathcal{H}_{\psi T} < 0$). In a similar model, Van der Ploeg & Withagen (1991) show that in their case, it is sufficient that the assimilative capacity or the social discount rate is high enough.

In the model presented here, the only capital stock is energy-related technology. Therefore, we could expect that a lower social discount rate would increase the

---

– 3.39 gives the general optimality conditions:

$$U'Y_F \;=\; -\gamma \int_t^\infty U'[C(s)]Y_P[F(s), P(s)]e^{-(\rho+\alpha)(s-t)}\,ds \qquad (3.45)$$

$$U'(C(t)) \;=\; \int_t^\infty U'[C(s)]Y_T[F(s), P(s)]e^{-(d_T+\rho(s-t))}\,ds \qquad (3.46)$$

These equations denote (cf. Equation 3.9) that at every point in time, the extra consumption due to an additional unit of fuel use (or due to one unit less invested) equals the present value of the loss in output (in utility terms) due to the enhanced concentration of GHGs due to this additional unit of fuel (or due to the last unit not spent on investment).

energy-related capital stock as well as environmental quality. However, as will be shown, the sign of any of the relevant variables is indeterminate. Using Cramer's rule, we have that:

$$\frac{dP}{d\rho} = \frac{\det J(P)}{\det J}$$

where $\det J$ is the determinant of the Jacobian matrix analysed above and $\det J(P)$ the determinant of matrix J with column vector $a = [-\psi - \xi \ 0 \ 0]'$ instead of the third column. Likewise,

$$\frac{dT}{d\rho} = \frac{\det J(T)}{\det J}$$

with $J(T)$ defined as the J with column vector a defined above instead of the fourth column. Note that $\det J > 0$ (see also Musu (1992)). However, both $\det J(P)$ and $\det J(T)$ are complex expressions with indeterminate sign. Moreover, the assimilative capacity and the social discount have both an indeterminate effect on $\det J(P)$ and $\det J(T)$, unlike in Van der Ploeg & Withagen (1991).

The main problem is that the derived demand for fossil fuel is positively related with T: $F_T > 0$, and hence: $\mathcal{H}_{\psi K} > 0$. This means that the sign of the effect of changes in ρ on P and T is ambiguous. The next subsection looks at specifications of the model where more patience in a society will lead to both more energy-related capital and a better environment for reasonable choices of the numerical values of the parameters.

Decentralised Version of the Model Optimisation models with a Central Planner are a convenient way of describing a more realistic economy. The decentralised version will be worked out in a model with a representative consumer and producer, and with a government.

producer The representative price-taking producer maximises current profits, π. These profits consist of the value of output[39] , $Y\{e(F,T),P\}$, minus the costs of renting the energy-related capital[40] ,rT, and the emission charges, τF. Formally, this can be described as:

$$\max_{F,T} \pi = Y\{e(F,T),P\} - rT - \tau F$$

---

[39]The price-level is set to unity for convenience.

[40]It is trivial to expand this model by including labour, L, in the production function: $Y\{e(F,T,L),P\}$ and with labour costs w per unit of labour. This would yield as additional first-order condition $Y_L = w$.

This gives as first-order conditions: $Y_F = \tau$ and $Y_T = r$. Below, it will be shown that these conditions coincide with the Central Planner social optimum for appropriate choice of the externality tax τ.

consumer The *representative price-taking consumer* is assumed to maximise intertemporal utility given his budget constraint[41]. This constraint states that his consumption can not exceed his income at any instant in time. Income[42] consists of profits, π, compensation of renting out energy-related capital, rT, and lump sum distribution of the emission charges, T. The consumer can spend his income on consumption, C, replacement investment, $d_T T$, and new investment in the energy-related capital stock, $\dot{T}$. Put differently, the decision of the consumer is to:

$$\max_C \int_0^\infty e^{-\rho t}[U(C)]dt \qquad\qquad \rho > 0$$

s.t.

$$C + \dot{T} + d_T T = \pi + rT + T$$

This gives first-order conditions $U'(C) = \xi$ and $\dot{\xi} = \xi(\rho + d_T - r)$.

government The government's sole goal is to levy Pigouvian taxes on the fossil fuel emissions of the producer and to redistribute them to the consumer. Hence:

$$T = \tau F$$

For each value of τ, there will be another equilibrium in this economy.

It is now straightforward to see that the first-order conditions of this decentralised economy coincide with those of the Central Planner model outlined above iff:

$$\tau = -\psi\frac{\gamma}{U'} \equiv \tau^* \qquad\qquad (3.47)$$

This means that the optimal Pigouvian tax, τ^*, equals the shadow value of environmental degradation due to the Greenhouse Effect in utility terms[43]. Note that

---

[41]Note that the budget constraint does not allow for lending and borrowing as a way to smooth consumption over time. Allowing for this possibility might change the outcomes quite dramatically, because fossil fuel use at a specific moment in time is de-linked from consumption in that period.

[42]If labour is included in the model, as hypothesised in the previous footnote, income is supplemented with the amount wL of labour income.

[43]This is the product of the costate variable of the optimal control problem and a fraction γ/U'. This fraction is to shift from 'fuel terms' to 'utility terms'.

this result is identical to that of the rudimentary economy at the beginning of this chapter.

As stressed, each level of emission charges τ gives a different steady state outcome. At the end of this section, the consequences of a too low level of carbon taxes are presented. It is shown that fossil fuel prices below their shadow price lead to an equilibrium with considerably lower levels of steady state consumption, technology and environmental quality. The conclusion is that, in the interest of future generations, a government should resist forces that try to keep the carbon tax at too low levels.

3.3.2   Numerical Results

In subsection 3.2.2, a specification was given for the rudimentary model of the beginning of this chapter. The same numerical model will be used here, with the difference that the energy function, $e(.)$, in $Y\{e(.), P\}$ is redefined to incorporate energy-related technology. Therefore, $e(F) = kF^\delta$ is replaced by a constant returns to scale energy function:

$$e(F, T) = F^\delta T^{1-\delta} \tag{3.48}$$

This means that, at the relevant part of the domain of the function, a doubling of both fossil fuel and energy technology doubles total energy input in production. However, this does not mean that production is doubled as well, because of the repercussion of higher energy use on the stock of GHGs, influencing production negatively via the Greenhouse damage function.

With the specification of equation 3.48, the Central Planner has the task to solve the following system:

$$\max_{C,F} \int_0^\infty e^{-\rho t} \frac{1}{1-\sigma} C^{1-\sigma} dt \quad \rho > 0$$

s.t.

$$
\begin{aligned}
C &= Y - I \\
Y &= F^\delta T^{1-\delta} \left(\frac{\bar{P}^2 - P^2}{\bar{P}^2} \right) \\
\dot{P} &= \gamma F - \alpha P \\
\dot{T} &= I - d_T T
\end{aligned}
$$

The same numerical values are taken for the parameters as in Section 3.2.2. The fossil fuel elasticity of production is fixed at $\delta = 0.3$. This implies that the energy function is $e(F, T) = F^{0.3} T^{0.7}$. With this specification, the optimal trajectory of the relevant variables can, in principle, be calculated[44,45]. In order analyse the results for policy purposes, the trajectories are discussed and a welfare analysis is given.

The trajectories of the relevant variables are depicted in Figure 3.8. The initial values $P(0)$ and $T(0)$ are chosen quite arbitrarily, but well below their steady state value. The most remarkable feature of the trajectories is the considerable overshooting of energy-related capital, T, vis-à-vis a smooth non-decreasing path to the steady state of P. This also holds for all other initial values, as long as $P(0)$ is not taken very close to P^{∞}[46]. Hence, the triple (T, C, ξ) changes much quicker than (P, F, ψ). This means that it is easier to change investment in energy-related capital then to adapt the level of fossil fuel use. This feature of the model is both due to the specification of the energy function (the parameter value δ) and due to the equation of the evolution of the environmental stock. However, the degree of overshooting appears not to be influenced by the size of α relative to the size of γ[47]. On the other hand, the degree of overshooting appears to be positively related to the size of δ[48]. This means that overshooting is reasonably robust to changes in the parameter values.

The idea that it is optimal to start off investing rapidly in energy-related capital at the expense of consumption has not been analysed in any of the other models discussed in Chapter 2. It is interesting to investigate this for numerical specifications of these models. This might give an idea whether the overshooting result holds generally for problems with abatement capital and pollution. In my opinion, the possible overshooting of the control and state variables is really worth analysing in depth. Especially interesting is the question what this means for policy recommendations on the basis of such results.

---

[44]The only new parameter is d_T, the depreciation rate of energy-related capital. This is set to $d_T = 0.1$.

[45]It is non trivial to do this for the numerical non-linear two-boundary point optimal control problem defined above. Standard shooting techniques in which appropriate initial values of the co-state variables are obtained via a trial-and-error method, are difficult in models with more than one state variable, as there is no obvious link between the 'trial' and the 'error'. Here, the programme APREDIC by Wymer (1992) is used to calculate solutions for this problem.

[46]At the extreme, if $P(0) = P^{\infty}$, and with $T(0) < T^{\infty}$, P will initially fall slightly (but with less than 1%) to return to the steady state value later on; T will slightly overshoot again.

[47]This was checked by changing the value of γ and putting the initial values of P_0 and T_0 at the same relative distance from their steady state values.

[48]From different scenarios calculated, it appears that an increase of γ with 10% leads to an increase in the degree of overshooting of some 5%.

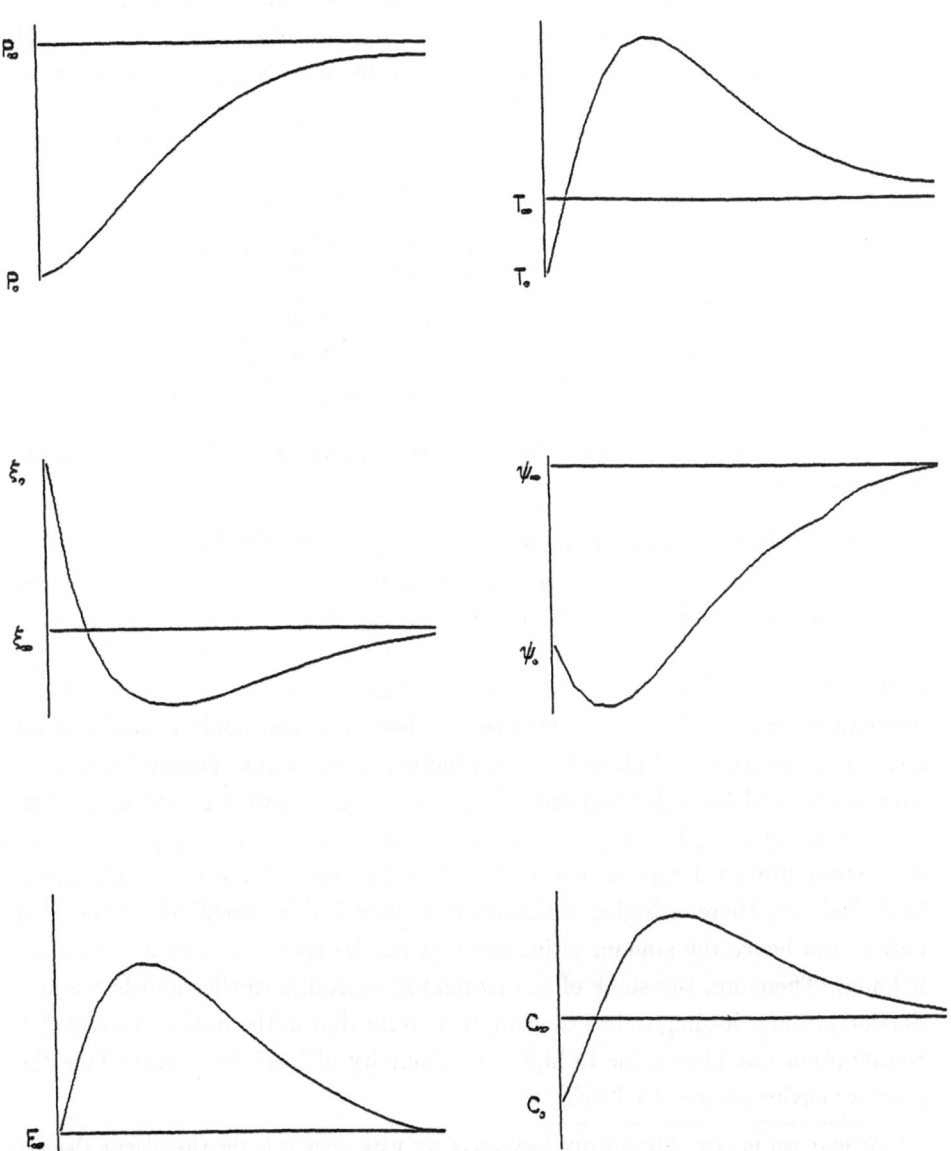

Figure 3.8: *Optimal trajectory for P, T, ψ, ξ, C and F for specification of the model given above*

Sensitivity Analysis of Parameters ρ, δ and γ   In Table 3.2, the steady state values of consumption (C), fossil fuel use (F), environmental degradation (P) and the energy-related capital stock (T) are given for different values of the parameters δ, ρ and γ[49,50]. The steady state levels of consumption for different specifications is given in Table 3.2. The benchmark case is Case 1 of the previous table (Table 3.1)[51].

case	δ	ρ	γ	P^∞	T^∞	F^∞	C^∞
1(bm)	0.30	0.03	0.50	715.68	178.95	7.16	15.33
2	0.30	0.01	0.50	557.09	770.99	5.57	44.06
3	0.30	0.06	0.50	813.03	30.24	8.13	3.89
4	0.40	0.03	0.50	763.76	39.17	7.64	4.57
5	0.30	0.03	1.00	715.68	89.47	3.58	7.67

Table 3.2: P^∞, T^∞, C^∞ and F^∞ for a variety of parameter values; Case 1 is the benchmark case.

As noted there, the absolute values of P^∞, T^∞, C^∞ and F^∞ do not have a direct interpretation, as they are not normalised in any way. Their importance lies in how they change due to changes in the parameters and how these values change over time. Comparing the benchmark case (bm) with Case 4, we see that a small change in the elasticity of production δ has a relatively modest impact on P and F, but a dramatic impact on T and C. The former is logical as one would expect that an increase in the fossil fuel elasticity of production tends to shift energy input away from capital and towards fossil fuel, which leads to more stress on the ecosystem. It is not trivial to explain why this change has such a enormous impact on T. The most straightforward explanation is that along the whole trajectory, with higher fossil fuel use, there is higher environmental degradation, negatively influencing output and hence the amount of income that can be spent on capital investment is lower. Therefore, the stock of energy-related capital is smaller and hence production is lower, leading to less consumption. Note that in the rudimentary model, consumption was higher due to higher productivity of fossil fuel outweighing the negative environmental feedbacks.

---

[49]We have not presented Sensitivity Analysis of the parameter $\hat{P}$ in the Greenhouse Damage function $f(P)$. A increase of $\hat{P}$ with 1% will increase C, F, P and T with 1% as well.

[50]Note that, given the specification of the utility function of the social planner as $U(C)$, the welfare evaluations can be made on the basis of consumption alone.

[51]As noted before, the value of the constant k in the energy function of the rudimentary model is chosen is such a way as to equalise the outcomes of the rudimentary model with those of the model presented in this section.

In Case 5, the percentage of fossil fuel use ending up in the atmosphere is doubled compared to the benchmark case. This decreases F, C and T by 50% and leaves P unchanged. As the elasticities in the energy function do not change, the optimal mix between F and T stays equal. With P unchanged, as in the rudimentary model, and with the optimal mix of F and T unaltered, consumption and investments will therefore drop by 50%.

The comparison the Case 1 (bm) with Case 2 and 3 shows the impact of changes of 'impatience' ρ on the economy. Note that changes in the steady state value of P due to variations in ρ do not differ from those given in Table 3.1 of the rudimentary model. At the same time, the changes in C are much more pronounced due to the dramatic impact that ρ has on the energy-related capital stock, T. The change pushes down output and consumption, though it does not have a feedback on P.

The direction of the change in P and in T due to a decrease in ρ could not be obtained unconditionally from the comparative statics exercise above. However, for the numerical specification of the model, a lower rate of time preference leads in the long run both to a higher energy-related capital stock and to more consumption. Besides, it gives rise to a better environment[52].

It seems that in the optimists versus pessimists debate in environmental economics (see Van der Ploeg & Withagen, 1991), this specification makes the pessimists' view more likely. The optimists would claim that by producing more, and investing part of this production in capital goods, there is a sounder base for future spending in environmental protection. However, with the specification of the model chosen above, producing more leads to more pollution which directly influences output. In this case, a more impatient economy will in this case burden future generations both economically and ecologically. Note, however, that the possibility of overshooting of the control and state variables gives a new dimension to the debate because optimists may now be right in the short run, while pessimists may be right in the long run and vice versa.

---

[52]As in the model without capital, giving up current consumption leads both to a better environment and to a higher level of future consumption. This reflects the idea that the traditional trade-off between consumption and environment is not valid in the long run. The results here support the idea that the fundamental trade-off is instead between current consumption and future consumption. As argued in Section 3.2.1, the specification with productive effects of pollution, responsible for the results, is chosen deliberately to highlight that importance of future repercussions of Climate Change on production. Put differently, if it were not ultimately for its impact on production, an elevated Greenhouse gas concentration would not be something to worry about.

Sensitivity Analysis on the Level of the Emission Charge τ In the decentralised version of the model, it was argued that there exists a level of emission charges, τ^*, set by the government, for which the economy will be at the social optimum. It may, however, well be that there are forces within the government that encourage an undervaluation of the environment. To make money, you have to spend money first and a myopic government with a time horizon not stretching too far beyond the next elections might not have an interest in optimal long term environmental protection. Especially, if this means taking unpopular decisions like increasing fossil fuel prices[53]. In Table 3.3, sensitivity analysis is presented on the effects of changes of the level of emission charges on the economic and ecological performance of the decentralised model.

case	ψ	τ	P^∞	T^∞	F^∞	C^∞
1(bm)	-1.41	1.39	715.68	178.95	7.16	15.33
2	-1.27	1.13	758.24	119.82	7.58	10.27

Table 3.3: *Sensitivity analysis on the effects of changes in emission charges*

In Case 2 of the table, the shadow price of pollution ψ is 10% lower than in Case 1, which corresponds to a emission charge 22.8% below the societal optimum value. This means a considerably lower steady state consumption and energy-related technology as well as a higher level of environmental degradation. It is instructive to see this comparison between 'correct' and 'too low' levels of emission charges in the restricted phase plane of (T, C)[54]. This is done in Figure 3.9. Assume that the economy is at the moment in the non-optimal steady state with 'too low' levels of externality taxes. In order to reach the optimal steady state, an 'eco-tax' has to be introduced that makes fossil fuel prices equal to their shadow prices. This leads initially to a drop in consumption (till the point where the $\dot{C} = 0$ line of the sub-optimal steady state hits the trajectory leading to the steady state with optimal charges). At a later stage, consumption levels would recover and go well beyond its initial level.

It is clear that a myopic government could have problems implementing this policy as the unpopular drop in consumption will occur in the short term whereas the beneficial rise in future consumption might take place only after the elections.

---

[53]In real life, a fairly convincing argument for these myopic forces in the government is that the shadow price of fossil fuel is not known with certainty and hence, it is not certain whether fossil fuel prices have to be increased or not.

[54]In the restricted phase plane, we have set P, ψ and F to their steady state values.

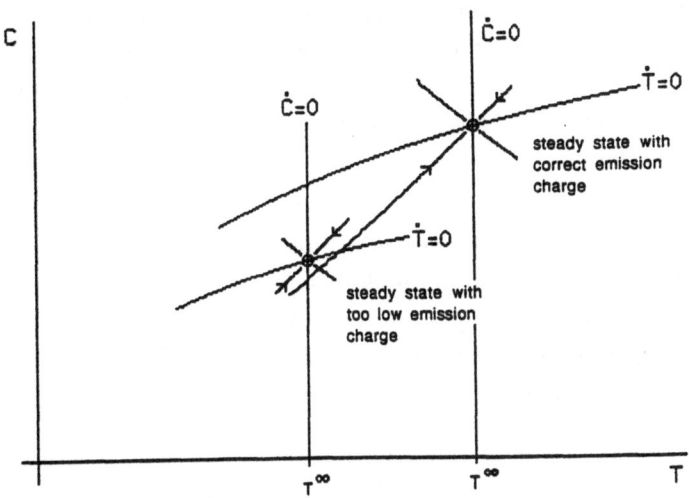

Figure 3.9: *Phase Plane of* (T, C) *with comparison between 'correct' emission charges and 'too low' emission charges*

Additional problems for the government might arise when human capital accumulation in the energy sector also needs government intervention in order to be on its optimal level. This issue is discussed in the next section.

3.4 Human Capital Accumulation

3.4.1 General Description

The previous section analysed the modelling of energy technology, T. It was stressed that T could be broadly defined, both including physical and human capital aspects of technology (K and H).

The aim of this section is to model the physical and human part of energy-related technology explicitly. This is worked out both for a general specification as well as for numerical examples. The trajectories of physical and human capital for different starting values of K and H are analysed.

Externalities in human capital formation are considered in order to analyse the potential problems for a market economy to reach a social optimum. Subsidies on human capital by a government are a possible way to achieve optimality. The externality of human capital formation is modeled such that production has decreasing

returns to scale without the subsidies and constant returns with the subsidies. This
could be an interesting first step to model endogenous growth in a polluted world[55].

Assume first, however, that externalities in human capital formation are absent.
In this case, the human and physical parts (H and K) of energy technology/capital
(T) are modeled as:

$$T = T(K, H) \tag{3.49}$$

Inclusion of K and H into the production function gives $Y\{e(F, K, H), P\}$. It is
assumed that $Y(.)$ is concave in all elements[56]. The stock of human capital (or
technological knowledge) changes over time through investment in human capital,
I_2 (see Gottinger, 1991) and through depreciation; physical capital evolves over
time in the standard way. This can be formalised as:

$$\dot{K} = I_1 - d_K K \tag{3.50}$$
$$\dot{H} = I_2^\theta - d_H H \qquad \theta \leq 1 \tag{3.51}$$

The fact that investment in human capital is likely to have decreasing returns to
scale (cf. Gottinger, 1991) is reflected by the fact that $\theta \leq 1$. The constant θ can
be referred to as a 'socially dependent transformation parameter of human capital
investment into knowledge'[57]. In this case, $\theta I_2^{\theta-1} - d_H$ is the effective net investment
in knowledge. Gottinger takes d_H equal to zero. Here it is assumed that $d_H > 0$,
but that $d_H < d_K$, reflecting the idea that physical capital depreciates at a faster
pace than human capital.

Combining these novel issues with the model of the last section gives the following
model:

$$\max_{C,F,I_2} \int_0^\infty [U(C)] dt \qquad \rho > 0 \tag{3.52}$$

s.t.

$$C = Y - I_1 - I_2 \tag{3.53}$$
$$Y = Y\{e(F, T), P\} \tag{3.54}$$

---

[55]One of the criticisms on the endogenous growth literature is its implicit assumption that
returns to capital are constant or increasing. It is often unclear on which facts this optimistic
view is based. One way to tackle this issue is by analyzing situations in which a government does
not intervene and one in which the government does intervene. The result of intervention is here
that constant returns to scale can be maintained.

[56]Furthermore, it is assumed that $Y_{FK} > 0$, $Y_{FH} > 0$ and $Y_{KH} > 0$, whereas $Y_{FP} < 0$, $Y_{KP} < 0$,
$Y_{HP} < 0$. These assumptions follow directly if we take Y$\{e(.),P\}$ as a product of $e(.)$ and $f(P)$,
as is done throughout in this text.

[57]Gottinger, 1991, p. 2

$$T = T(K, H) \tag{3.55}$$
$$\dot{P} = \gamma F - \alpha P \tag{3.56}$$
$$\dot{K} = I_1 - d_K K \tag{3.57}$$
$$\dot{H} = I_2^\theta - d_H H \qquad \theta \leq 1 \tag{3.58}$$

The necessary conditions for a solution of the model using Pontryagin's Maximum Principle are:

There exist co-state functions ψ for environmental degradation, ξ for physical capital accumulation and χ for human capital accumulation, such that[58]:

$$
\begin{aligned}
\mathcal{H}(C, F, I_2, P, K, H, \psi, \xi, \chi) &= U(C) + \psi(\gamma F - \alpha P) + \chi[I_2^\theta - d_H H] + \\
&\quad + \xi[Y\{e(F, K, H), P\} - C - I_2 - d_K K]
\end{aligned}
$$

With:

$$\frac{\partial \mathcal{H}}{\partial C} = U'(C) - \xi = 0 \tag{3.59}$$

$$\frac{\partial \mathcal{H}}{\partial F} = \gamma \psi + \xi Y_F = 0 \tag{3.60}$$

$$\frac{\partial \mathcal{H}}{\partial I_2} = -\xi + \theta \chi I_2^{\theta-1} = 0 \quad . \tag{3.61}$$

$$\frac{\partial \mathcal{H}}{\partial P} = \xi Y_P - \psi \alpha \quad \text{or} \quad \dot{\psi} = (\rho + \alpha)\psi - \xi Y_P \tag{3.62}$$

$$\frac{\partial \mathcal{H}}{\partial K} = \xi Y_K - \xi d_K \quad \text{or} \quad \dot{\xi} = (\rho + d_K)\xi - \xi Y_K \tag{3.63}$$

$$\frac{\partial \mathcal{H}}{\partial H} = \xi Y_H - \chi d_H \quad \text{or} \quad \dot{\chi} = (\rho + d_H)\chi - \xi Y_H \tag{3.64}$$

It is assumed, as before, that the limit conditions hold, so that an analysis of the internal solution(s) of the model suffices. Note that in Equation 3.61, $\xi = \chi$ if $\theta = 1$. This means that the shadow value of K and H are equal if the transformation parameter of human capital investment into knowledge is one. For other values of θ, this first order condition says that ξ and χ are equal in effective terms. The other equations above are similar to the ones discussed in section 3.3.1 and need no further elaborating.

The derived demand functions for C, F and I_2 can be calculated by taking the derivatives of $\mathcal{H}_C$, $\mathcal{H}_F$ and $\mathcal{H}_{I_2}$ with respect to C, F, I_2, P, K, H, ψ, ξ and χ. This

---

[58] For convenience, the derivatives $Y_e e_F$, $Y_e e_K$ and $Y_e e_H$ will be written as Y_F, Y_K and Y_H respectively. Note also that I_1 is substituted out of the system. The transversality conditions are not explicitly stated (see Section 3.3.1).

gives:

$$C(\xi) \qquad\qquad \text{with} \quad C_\xi < 0$$
$$F(P,K,H,\psi,\xi) \quad \text{with} \quad F_P < 0\,,\, F_K > 0\,,\, F_H > 0\,,\, F_\psi > 0 \text{ and } F_\xi > 0$$
$$I_2(\xi,\chi) \qquad\quad \text{with} \quad F_\xi < 0\,,\, F_\chi > 0$$

These derived demand functions can be substituted into the differential equations for the state and co-state equations. For general non-linear specifications, little is known about the stability properties of optimal control systems of three or more state variables. Therefore, we will concentrate on the restricted differential equation system where P and ψ are fixed $(\bar{P}, \bar{\psi})$[59]:

$$
\begin{aligned}
\dot{K} &= Y[e\{F(\bar{P},K,H,\bar{\psi},\xi),K,H\},\bar{P}] - C(\xi) - I_2(\xi,\chi) - d_T T \\
\dot{H} &= I_2^\theta(\xi,\chi) - d_H H \\
\dot{\xi} &= (\rho + d_K)\xi - \xi Y_K[e\{F(\bar{P},K,H,\bar{\psi},\xi),K,H\},\bar{P}] \\
\dot{\chi} &= (\rho + d_H)\chi - \xi Y_H[e\{F(\bar{P},K,H,\bar{\psi},\xi),K,H\},\bar{P}]
\end{aligned}
\tag{3.65}
$$

The Jacobian matrix of the linearised restricted differential equation system, evaluated at the steady state $(K^\infty, K^\infty, \xi^\infty, \chi^\infty)$ can be derived to investigate the local stability characteristics. It is shown in Appendix A.2 that the equilibrium solution of the restricted dynamical system 3.65 is a saddle point if the rate of time preference ρ is sufficiently low and the steady state is locally asymptotically stable (cf. Section 3.3.1).

Interpretation of Conditions of Steady State At the steady state[60] of the system described above, the following conditions hold[61]:

$$\gamma F^\infty = \alpha P^\infty \tag{3.68}$$

---

[59]Note that in Section 3.2.1 , focus was on the state variable P. In Section 3.3.1, we concentrated on the dynamics around the state variables P and T. In this section, we have $T = T(K,H)$. Hence, in the absence of possibilities to look at the dynamics of P, K and H at the same time, the most relevant dynamics to consider is that between K and H.

[60]Note that, the only steady state analysed is the one with zero long-term growth (the stationary state). In contrast, in the endogenous growth literature, constant (non-zero) growth is primarily analysed.

[61]The general optimality conditions similar to the ones given in Section 3.3.1. Here, besides Equations 3.45 and as substitute for Equation 3.46, the conditions are that:

$$U'(C(t)) = \int_t^\infty U'[C(s)]Y_K[s]e^{-(d_K+\rho(s-t))}\,ds \tag{3.66}$$

$$U'(C(t)) = \theta I_2^{\theta-1}\int_t^\infty U'[C(s)]Y_H[s]e^{-(d_H+\rho(s-t))}\,ds \tag{3.67}$$

$$Y^\infty = C^\infty + I_2^\infty + d_K K^\infty \tag{3.69}$$

$$I_2^{\theta\infty} = d_H H^\infty \tag{3.70}$$

$$\frac{1}{\gamma}U'(C^\infty)Y_F = \frac{1}{\rho+\alpha}U'(C^\infty)Y_P \tag{3.71}$$

$$\rho + d_K = Y_K\{e(F^\infty, K^\infty, H^\infty), P^\infty\} \tag{3.72}$$

$$\rho + d_H = Y_H\{e(F^\infty, K^\infty, H^\infty), P^\infty\}\theta I_2^{(\theta-1)\infty} \tag{3.73}$$

Equations 3.68, 3.69, 3.71 and 3.72, are the same as Equations 3.41, 3.42, 3.43 and 3.44 and need no further explanation[62]. Equation 3.70, states that in the steady state, investment equals depreciation, hence there is no net investment. Equation 3.73, states that there is no change in the valuation of the stock of human capital. Therefore, the marginal product of this capital in effective units equals the sum of social discount rate and depreciation rate. Note that the right hand side of Equation 3.73 is $\frac{\partial Y}{\partial H}\frac{\partial H}{\partial I_2}$, where the latter term is one for θ being equal to unity.

Decentralisation Similarly as in Section 3.3.1, the benevolent dictator optimal control model can be rewritten in a decentralised form to show that the social optimum can be supported by a market economy. For simplicity, we restrict the analysis to the case that the parameter of transformation of human capital investment into knowledge θ is equal to one. In this case, the decision making problem of the representative producer is:

$$\max_{F,K,H} \pi = Y\{e(F,K,H),P\} - r_K K - r_H H - \tau F$$

yielding first order conditions $Y_F = \tau$, $Y_K = r_K$ and $Y_H = r_H$. The representative consumer maximises intertemporal consumption over time with the budget constraint:

$$C + \dot{K} + d_K K + \dot{H} + d_H H = \pi + r_K K + r_H H + T$$

The government sets pollution charges τ and redistributes the amount $T = \tau F$ to the consumers. As before, for each level of τ, another equilibrium will result and the government can guarantee a socially optimum outcome by putting the emission tax equal to the shadow price of the environment. This issue of decentralisation

---

These equations denote (cf. Equation 3.46) that at every point in time, the extra consumption due to one unit less invested (in either human or physical capital) equals the present value of the loss in output (in utility terms) due to the enhanced concentration of GHGs due to this additional unit of fuel (or due to the last unit not spent on investment).

[62]Note that Equation 3.69 has, compared to Equation 3.42, the additional variable I_2

case	β	θ	P^∞	K^∞	H^∞	T^∞	C^∞
1(bm)	0.60	0.95	715.68	85.45	71.02	178.95	15.33
2	0.80	0.95	715.68	155.15	49.30	278.17	20.14
3	0.60	0.90	715.68	75.68	55.23	150.45	13.89

Table 3.4: P^∞, K^∞, H^∞, T^∞ and C^∞ for a variety of parameter values; Case 1 is the benchmark case

will be re-addressed at a later stage when a public good aspect of human capital investment is analysed. In that case, a government has the additional task of subsidising investment in knowledge in order to guarantee Pareto optimality.

3.4.2   Numerical Results

In order to calculate the trajectories of the relevant variables for welfare evaluations and sensitivity analysis, the functional forms of the functions need to be specified. This has been done exactly as in Subsection 3.3.2. Additionally the function of energy-related capital is specified as a constant returns to scale Cobb-Douglas function:

$$T = \kappa K^\beta H^{1-\beta} \tag{3.74}$$

Hence, the production function is specified as:

$$Y\{e(F,K,H),P\} = F^\delta \kappa^{1-\delta} K^{(1-\delta)\beta} H^{(1-\delta)1-\beta} f(P) \tag{3.75}$$

where $f(P)$ is defined as above.

Equation 3.75 states that the energy function has also constant returns to scale. The evolution of the capital stocks K and H is as given in Equations 3.50 and 3.51.

The parameter values are the same as in Subsection 3.3.2. Additionally, the elasticity in the CRS Cobb-Douglas technology function, β, is set to 0.6, though results of sensitivity analysis will be presented for a higher value as well ($\beta = 0.8$). The value of θ in the investment function of human capital is set close to one at 0.95, with robustness analysis for $\theta = 0.90$. The parameter values of κ, d_K and d_H are chosen in such as way that the results of the benchmark case coincide with those of Table 3.2 of Section 3.3.2[63].

In Table 3.4, the steady state results for the values discussed above are presented. Note that P^∞ is the same for all parameter variations. The reason is that in the

---

[63]This is $\kappa = 2.255$, $d_K = 0.133$ and $d_H = 0.083$.

steady state the equations of motion of P and ψ form an independent subsystem in P and F[64].

A change in β has quite a considerable impact on the steady state results, as is seen from comparing Case 1 and Case 2. This means that the results we have obtained are not robust with respect to the choice of β. It also means that it is very important to try to get good real life data on the stock of physical versus human capital in the energy-related technology.

A lower value of the effectiveness of human capital investment, expressed by the coefficient θ, means obviously that human capital is substituted for physical capital. In fact, the ratio of human to physical capital shifts from 0.83 to 0.73 in Table 3.4. Note that has been kept d_H fixed, so that the economy as a whole is worse off.

Optimal Trajectories In order to focus on the evolution over time of human and physical capital (H and K), the restricted differential system (with the pollution part of the model fixed at its steady state level: $P = P^\infty$ and $\psi = \psi^\infty$) is solved for different initial values of H_0 and K_0. The computer programme used to calculate the optimal trajectory of K_t and H_t is GAMS[65,66,67]. In Figure 3.10, these trajectories are presented for K and H. In Graph b of this figure, the starting values K_0 and H_0 are taken equiproportionally below their steady state values. This results in a path for both K and H that is monotonically increasing over time.

In Graph a of the same figure, K_0 is taken relatively low compared to the initial value of human capital, H_0. In that case, investments are primarily directed to increases in the physical capital. This leads the stock of human capital to fall temporarily, with a rapid recovery of K. After this initial adjustment phase, both

---

[64]This becomes clear when ξ is substituted out of Equation 3.56 using Equation 3.60. Alternatively, see Equations 3.68 and 3.71 and note that $\frac{Y_F}{Y_F}$ is independent of K and H given our specification of the production function in Equation 3.75.

[65]GAMS is a computer programme developed by the World Bank, among other things to analyse systems of difference equations with a finite horizon. By increasing the horizon up to the moment that additional periods do not alter the time path of the variables any more, the infinite time solution can be approximated. In the last periods, the time-path starts to behave erratically. This is in accordance with the Turnpike Theorem.

[66]For reasons unclear to me and to Wymer himself, the computer programme APREDIC by Wymer (that was used in the previous section for the calculations of the trajectories) was not able to solve the restricted dynamic equation system.

[67]In Cesar (1993-b), two-dimensional slices of the model are analysed (esp. (K, ξ) and (H, χ)). These are similar to the phase planes from Ramsey growth models.

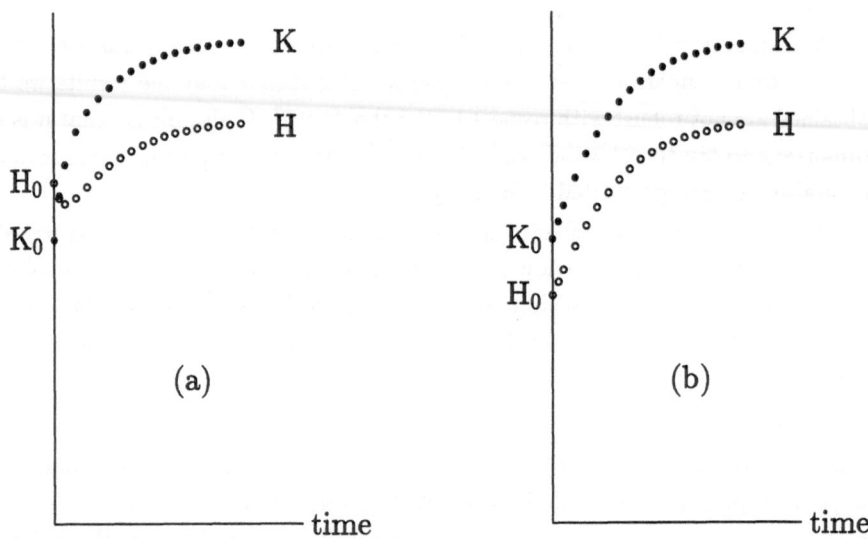

Figure 3.10: *Trajectory of K and H for the Cases (a) and (b)*

types of capital growth monotonicly over time, as in Graph a. The reverse pattern would occur when H_0 is relatively low compared to K_0. This pattern is well known from two-sector growth models.

In real life, it might be the case that the level of human capital is relatively low compared to that of physical capital. This may be due to externalities involved in knowledge formation, as analysed below. In that case, it would be optimal to concentrate investment initially on human capital accumulation.

3.4.3   The Public Good Character of Human Capital

As shown above, a decentralised economy with appropriate emission charges (to capture the pollution externality) is capable of mimicking the socially optimal Central Planner's results, in the presence of both human and physical capital.

However, as Grubb (1989) notes, "energy technology development is such a big and risky business that private companies are often reluctant to undertake it alone, and there are many other hurdles to the development of new technologies" (Grubb, 1989, p.30).

The aim of this section is to capture this additional market failure and to see what type of economic policy is needed in society to obtain a social optimum. In order to analyse this question, technology is redefined as:

$$T^A = T(K, H, H_A) \tag{3.76}$$

where H_A is the average level of human capital in the society (Lucas, 1988). The intuition behind this specification is that technology of any firm depends on human and physical capital of that firm as well as on the average level of human capital in the society at large. Investment in human capital is therefore assumed to both increase H and H_A[68]. Following Lucas (1988), human capital is modeled as[69]:

$$T^A(K, H, H_A) = \kappa K^\beta H^{(1-\beta)\epsilon} H_A^{(1-\beta)(1-\epsilon)} \tag{3.77}$$

Typically, economic agents would assume H_A to be given so that $T^A(.)$ will have decreasing returns to scale in the absence of appropriate government intervention. Subsidising H optimally could carry the economy back to a constant returns to scale situation.

This is crucially important for endogenous growth models, where an explanation of why constant returns can be assumed is often lacking.

In order to analyse this, the first order conditions for the representative producer are compared for different cases. First, in the decentralised version of the model of Section 3.3.1, these are:

$$
\begin{aligned}
Y_F &= & \delta Y &= \tau \\
Y_K &= & (1-\delta)\beta Y &= r_K \\
Y_H &= & (1-\delta)(1-\beta)Y &= r_H
\end{aligned}
\tag{3.78}
$$

In the presence of externality aspects of human capital, as defined by $T^A(K, H, H_A)$ above, the corresponding first order conditions are[70]:

$$
\begin{aligned}
Y_F^A &= & \delta Y &= \tau \\
Y_K^A &= & (1-\delta)\beta Y &= r_K \\
Y_H^A &= & (1-\delta)(1-\beta)\epsilon Y &= r_H
\end{aligned}
\tag{3.80}
$$

---

[68]A patent market is assumed not to exist.

[69]Lucas (1988) takes K fixed in order to concentrate solely on the dynamics of H and H_A alone. Here, K will not be fixed. However, concentration is on the steady state only.

[70]Y^A is defined as:

$$Y^A = F^\delta \{\kappa K^\beta H^{(1-\beta)\epsilon} H_A^{(1-\beta)(1-\epsilon)}\}^{1-\delta} f(P) : \tag{3.79}$$

case	ϵ	P^∞	K^∞	H^∞	T^∞	C^∞
1(bm)	1.00	715.68	85.45	71.02	178.95	15.33
2	0.90	715.68	78.15	59.03	157.52	14.62
3	0.80	715.68	70.73	48.01	136.60	13.77

Table 3.5: P^∞, K^∞, H^∞, T^∞ and C^∞ with the introduction of an externality indicated with α; Case 1 is the benchmark case

This means that, with the externality, relatively too little human capital is used in equilibrium, as the rate of return on H has not changed with the lowering of marginal productivity of H. In Table 3.5, the equilibrium levels for the relevant variables are shown for different values of ϵ. In the benchmark case, ϵ equals unity and hence, there is no externality[71]. In Cases 2 and 3, the decentralised economy uses too little human capital. This means that the overall economy is less efficient, with lower steady state consumption. Note also that in equilibrium, there is less physical capital. This means that, for the specification of $T^A(.)$ above, the substitution effect towards physical capital is outweighed by the 'efficiency' effect.

Assume that a subsidy by the government on human capital investment of the following form is introduced:

$$S = sH \qquad (3.81)$$

where S is the lump sum taxation for the representative consumer in order to finance the subsidy and s is the subsidy per unit of human capital.

The decision making process of the representative firm with the subsidy can be formalised as:

$$\max_{F,K,H} \pi = Y\{e(F,T(K,H,H_A)),P\} - r_K K - r_H H - \tau F + sH$$

With the specification of the production function given in Equation 3.79, the first order conditions of the representative producer become:

$$
\begin{aligned}
Y_F^A &= & \delta Y &= \tau \\
Y_K^A &= (1-\delta)\beta Y &= r_K \\
Y_H^A &= (1-\delta)(1-\beta)\epsilon Y &= r_H - s
\end{aligned}
\qquad (3.82)
$$

For each value of s and τ, another equilibrium outcome results. The system of

---

[71] The values of the relevant variables are thus equal to the benchmark case in Table 3.4.

Equations 3.82 equals the corresponding first order conditions of 3.78 iff[72]:

$$
\begin{aligned}
s &= r_H(1 - \epsilon) \equiv s^\star \\
\tau &= -\psi \frac{\gamma}{U'} \equiv \tau^\star
\end{aligned}
\tag{3.83}
$$

Note that with subsidies at rate $s^\star$, the effective return to human capital is αr_H instead of r_H. This gives, on the margin, the correct incentive for the representative producer to weigh physical and human capital optimally. Therefore, the Pareto optimal steady state can still be attained with the appropriate measures by the government.

3.5   Summary and Conclusions

In this chapter, various models of the Greenhouse Effect have been developed. First, a pollution model with one state variable (the stock of atmospheric Greenhouse gases) and without abatement possibilities was worked out. Pollution was taken to affect production possibilities rather than utility directly. This specification implies that the fundamental intertemporal trade-off is between current consumption and future consumption rather than between current consumption and future pollution. Also, higher time-preference was proven to lead to both lower future consumption and more pollution, as long as the negative impact of the Greenhouse Effect on production is high enough.

Secondly, steady state equilibria were compared for different non-linear forms of the natural assimilation function. It appeared that the results were highly sensitive to small variations in the parameters. Given the large uncertainty with respect to the actual assimilation of Greenhouse gases for concentration levels above the current ones, explicit modelling of this uncertainty in future research is therefore of the utmost importance.

Thirdly, Ramsey models with a pollution stock were introduced, where output was optimally allocated intertemporally between consumption, fossil fuel and energy-related investment. Conditions under which the steady state is locally asymptotically stable were analysed and comparative statics results for changes in the time preference were presented. For the numerical specification of the model, a higher discount rate was shown to imply a higher level of consumption, abatement capital and environmental quality in the long run. Interestingly, the trajectories of,

---

[72]The optimal value of emission charges, $\tau^\star$, was calculated in Section 3.3.1.

among other variables, consumption and energy-related capital showed a considerable overshooting. It is up to future research to see how general such overshooting is in Ramsey models with a pollution stock.

Fourthly, energy-related capital was split in its human and physical part and conditions were analysed for locally asymptotical stability of the restricted dynamic system where pollution was kept at its steady state level. Trajectories for human and physical capital were discussed for different starting values of the state variables. For a relatively low initial level of human capital compared to physical capital, a rapid accumulation in the former was shown to take place at the expense of physical capital, which is in line with results in standard two-sector growth models.

Finally, spill-overs in the accumulation of human capital were assumed to be present, where the externalities led to decreasing returns to scale in production in the absence of government intervention. With internalisation, production exhibited constant returns to scale so that, sustained growth depended crucially on economic policy. The externalities in pollution and knowledge were internalised through carbon taxes and human capital subsidies respectively. This issue of whether long-term growth at sustainable pollution levels is possible, will, in my opinion, prove to be one of the most challenging research topics of the coming years.

The discussion ends this chapter. Next, a new externality will be introduced that results from the transboundary character of the Greenhouse Effect.

Chapter 4

Multi Country Modelling of the Greenhouse Effect

Non può, pertanto, uno signore prudente, né debbe osservare la fede,
quando tale osservanzia li torni contro e che sono spente le cagioni che la
feciono promettere. E, se gli uomini fussino tutti buoni, questo precetto
non sarebbe buono; ma, perché sono tristi e non la osservarebbono a te,
tu etiam non l'hai ad osservare a loro.[1]

N. Machiavelli (1513), Il Principe.

4.1 Introduction

Over the last few decades, a dramatic shift has occurred in the scale of environmental problems, from the local and regional level to a fluvial (Rhine, Baltic Sea, Mediterranean, etc.), continental (acid rain) and global (Greenhouse Effect, ozone depletion) level. The common feature of these latter three scales is that the pollution is transboundary in that emissions (discharges) in one country affect the environmental quality in other countries.

Isolated national policies with respect to emission reductions are likely to overlook these spill-over effects arising from the transboundary nature of the pollution problem. Therefore, transboundary pollution problems obviously call for interna-

---

[1]Edition of Rizzoli, 1990, p.156; English translation (G. Bull; Pinguin Classics, 1981, p. 99-100): *"So it follows that a prudent ruler cannot, and must not, honour his word when it places him at a disadvantage and when the reasons for which he made his promise no longer exist. If all men were good, this precept would not be good; but because men are wretched creatures who would not keep their word to you, you need not keep your word to them."*

tional environmental policies. This is because one can reasonably expect that there are economic gains to be made when international dependencies across countries are present.

In economic terms, the spill-over effects form an externality that is not appropriately valued. Correcting for the resulting inefficiency is, in principle, possible through international cooperation aimed at internalising this externality.

This spill-over issue is also central in the literature on international macroeconomic policy coordination, as it has developed over the last dozen years (for a recent overview, see Currie et al. (1989)). To a large extent, this literature has followed a game-theoretic approach to describe the challenges of maintaining cooperation vis-à-vis the temptations of, for instance, indulging in beggar-thy-neighbour policies. This gaming approach seems to provide an appropriate framework as there are typically several interacting agents involved, having both common and opposing interests with each having some influence on the final outcome. Both static and dynamic (e.g. differential) games have been used in this context. It is assumed that the reader is familiar with the main concepts of game theory, although the essential features of differential game theory will be briefly reviewed in this chapter (**Section 2**), with special attention being given to the distinction between open-loop and feedback strategies.

Following the introduction of game theory in the literature on international macroeconomic policy coordination, environmental economists have also started to use the same tools over the last few years[2]. Examples include Mäler (1989, 1991), Kaitala, Pohjola & Tahvonen (1990) and Cesar (1989) on acid rain; Musu (1991) on water pollution in the Mediterranean basin and Barrett (1990), Cesar (1990, 1993-a), Hoel (1990), van der Ploeg & de Zeeuw (1990)[3] and Tolwinski (1992) on the Greenhouse Effect[4,5]. Most papers use differential game theory, where the

---

[2]In natural resource economics, this gaming framework has been used before, e.g. Levhari & Mirman (1980) in a fishery game and Reinganum & Stokey (1985) in a game of extraction of exhaustible resources.

[3]The article by van der Ploeg & de Zeeuw (1992) does not speak explicitly about the Greenhouse Effect, but the authors do model a type of international pollution with a symmetric transportation matrix, such as the Greenhouse Effect.

[4]For additional papers, see also a recent Conference volume on dynamic games with special attention to international environmental problems (Conches-Grimentz, 1992).

[5]From a modelling point a view, a main difference between the various forms of pollution is that Greenhouse gases disperse equally around the globe, so that one stock variable suffices to describe the GHG concentration at each point on earth. For most other pollution problems, including acid rain and water pollution, the transportation of emissions (or discharges) through

pollution stock(s) is (are) the only state variable(s). Exceptions are the papers by Mäler (1989) and Barrett (OECD, Ch. 3, 1991), that have a static game setting. Conclusions from these papers are that optimal cooperation leads to abatement efforts such that marginal emission-reduction costs are internationally equalised[6]. It is also becoming generally understood that a cooperative global programme to reduce Greenhouse gas emissions has very considerable cost advantages over isolated national strategies[7]. For a discussion of the papers, see Section 2.3.1

In **Section 3** of this chapter, our earlier analysis will be extended by looking at strategic issues raised by the international nature of the Greenhouse Effect in the presence of a capital stock[8]. This will widen the policy options available to a single government in choosing an appropriate policy response and presumably make the analysis more realistic and relevant.

The major conclusion of the numerical calculations in this section is that co-operation leads to substantially lower levels of fossil fuel use and hence of the atmospheric concentration of GHGs. The effects of concerted action on technology is quite marginal, the size depending on the substitution possibilities between fossil fuel and energy technology.

Furthermore, the feedback case corresponds to a lower steady state level of social welfare than the open-loop case in our model. This implies that the potential gains from cooperation are even higher when the feedback rather than the open-loop

---

air (or water) is asymmetric, because of prevailing wind (or stream) directions. An extreme example of this asymmetry in a static framework is up-stream versus down-stream river pollution. The consequence of this asymmetry is that, for most environmental problems, the deposition of pollution varies per region so that different stock variables are needed for the description of the concentration of pollution in different areas. This makes a formal game-theoretic analysis more complicated.

[6]This holds for the Greenhouse Effect. For asymmetric pollution problems (see previous footnote) the result is a bit more complicated, as marginal emission reduction costs are, in the social optimum, more elevated in the vicinity of high impact regions than away from these regions. For instance, for acid rain in Europe, marginal abatement costs in Czechoslovakia are higher, in the Pareto optimum, than in Spain. See Mäler (1989) for the optimality conditions.

[7]Barrett (OECD, Ch. 3, 1991) and Nordhaus (1990) present some tentative estimates on potential cost savings in the case of the Greenhouse Effect. On the Acid rain problem in Europe, quite a few estimations for cost-reduction through international cooperation in the case of Acid Rain in Europe have been published. For an overview, see Cesar (1989). The most well-known estimate is by Mäler (1989) who assessed the gains of cooperation from a pan-European SO_2 reduction scheme to be 6.3 billion DM annually.

[8]Xepapadeas (1991) and Van der Ploeg & Ligthart (1993) describe an international pollution game where both the pollution stock and the stock of 'technology' have a public good character.

solution is taken as a basis for comparison. The reason is that in the feedback case, countries anticipate that increases of the emission in one country push the optimal level of emissions in the other countries down. Hence, the marginal benefits to the environment of an additional unit of emission reduction are lower in the feedback case than they would be in the open-loop Nash equilibrium[9].

Given the predicted benefits of cooperation, the next logical question is why cooperation is difficult to achieve in real life. One factor is the incentive problem that, given cooperative behaviour of the other countries, any player typically gains from reneging. In **Section 4**, we focus on how to avoid this free rider problem, and how to preserve cooperation with self-enforcing agreements in differential games, using trigger and renegotiation-proof strategies[10,11].

It will be proven that cooperation can be sustained in both cases, as long as cooperation is in the interest of all the negotiating parties. The question is how to translate these abstract conclusions into policy recommmmendations for environmental negotiations. In my opinion, the results, though not directly applicable, do highlight some of the essential issues, that need to be dealt with in International Environmental Agreements. Examples include the issue of how to punish cheaters in a self-enforcing way and how to deal with the question of monitoring compliance.

Existing International Environmental Agreements show a whole range of possible ways to monitor compliance, from reporting by authorised international officials to self-reporting[12]. Punishment of non-compliance is, given the inadequacy of international law to enforce agreements, a difficult issue and most treaties do not

---

[9]Note that in a case **with** uncertainty and with **only one** actor, the feedback solution is generally better than the open-loop solution. The reason is that the actor can adapt when the circumstances require that.

[10]In Chapter 5, the problem that some countries might be worse off with than without cooperation will be addressed (allegedly Canada and Russia in Greenhouse negotiations).

[11]The avoidance of free riding in international environmental agreements has also been considered in a quite different context, by looking at the stability issue of coalitions of cooperating countries (See for instance Barrett (1989), Carraro & Siniscalco (1991), Heal (1991) and Parson & Zeckhauser (1992).

[12]For instance, the Montreal Protocol (ozone) and the Geneva Convention (sulphur dioxide) oblige members to report annually on the substances in question. At the same time, these agreements do not allow for observers from other countries. The weakness of such self-monitoring from an incentive point of view are obvious. On the other hand, the North Pacific Fur Seal Treaty allows authorised officials of each signatory to board and search vessels of other signatories. And the Antarctic Treaty allows all signatories the right to designate observers to carry out inspections with complete freedom of access at any time or to any area of Antarctica (Barrett-OECD, Ch. 3, p.129, 1991).

explicitly deal with this. An exception is the North Pacific Fur Seal Treaty that states that in the event of violation, and provided signatories cannot agree to remedial measures, any signatory may give notice of its intentions to terminate the agreement. The consequence of this notification is that the agreement will be terminated nine months later (Barrett-OECD, Ch. 3, p.129, 1991).

The game-theoretic approach taken in this chapter on how to sustain cooperation could give insight in analyzing these and other real world issues in international (environmental) negotiations[13].

4.2 Differential Game Theory

In the previous chapter, optimal control methods were used in models with one decision maker (the Central Planner). Differential game theory looks at the strategic aspects arising in a multi-player context, where each of them is described in an optimal control framework. Başar & Olsder (1982; p. 210) define differential games as continuous time infinite dynamic games wherein the evolution of the state is described by a differential equation and wherein the players act throughout the time interval[14].

These techniques were introduced by Isaacs in 1954 and mostly developed in the engineering literature in a zero-sum context, especially for military pursuit modelling[15]. In economics, however, players generally have both common and opposite interests and therefore non-zero sum differential games seem much more appropriate (de Zeeuw, 1991)[16].

Examples of such games in the economics literature are Reinganum & Stokey (1985) (resource economics), Miller & Salmon (1985) (monetary policy), Hanig

---

[13]In Chapter 5, additional issues will be dealt with that occur if first-best (cooperative) solutions are not feasible because one of the negotiating parties would lose relative to non-cooperation. Especially technology transfers and the linkage of different negotiations will be explored.

[14]Likewise, difference games are discrete time infinite dynamic games with difference equations governing their evolution over time. The term 'dynamic' is used to distinguish these games from 'repeated' games where players are involved in a repetition of the original game over time. In dynamic games, the game itself changes over time as the state evolves.

[15]In pursuits, the zero-sum idea makes sense. A Scud missile, for instance, and a Patriot anti-missile missile are fully antagonistic in the sense that they have no common interests. Likewise, pilots in a dog-fight typically only have opposite interests (though they would probably both be better off if they make peace).

[16]The common interest in the case of a global pollution game is to curb emissions; the opposing interest is that each prefers the other to carry out the abatement.

(1986) (industrial economics), and many others.

In what follows, a brief introduction to the theory of differential games is given. This is mainly based on Başar & Olsder (1982), Hanig (1986), Fudenberg & Tirole(1991) and de Zeeuw (1991). The concentration is on two-person games.

In a two-person differential game, both players affect the evolution of a **state variable** x and the **equations of motion** are defined by the differential equation system[17]:

$$\dot{x}(t) = g(x(t), u_1(t), u_2(t), t) \qquad\qquad x(0) = x_0$$

In this equation, $u_1(t)$, $u_2(t)$ are the **control variables** of player 1 and player 2 respectively. It is assumed that $u_i(t)$ is a compact continuous function of time and the functions $g(.)$ are taken to be continuously differentiable.

Each player i $(i = 1, 2)$ chooses a control variable u_i so as to maximise his **objective function**[18]:

$$J_i(u_1(t), u_2(t)) = \int_0^T f_i(x(t), u_1(t), u_2(t), t)dt + Q_i(x(T))$$

where $Q_i(x(T))$ is the scrap value at time T.
In the applications in this chapter, infinite time problems will be considered where the scrap value is zero (i.e. $Q_i(x(\infty)) = 0$). The general formulation with $T \leq \infty$ is taken here in the overview section. The functions $f_i(.)$ and $Q_i(.)$ are assumed to be continuously differentiable.

In order to define the players' strategies, the type of information available to them has first to be specified. The **information structure** is defined as a set valued function $\eta_i(t) = \{x(s), s \,\epsilon\, S(t)\}$, where $S(t)$ is a subset of $[0, T]$, and $S(t)$ is non-decreasing in t. Başar & Olsder (1982) define five types of information structures. The two types that are relevant for the analysis here are[19]:

open-loop: This information structure can be formalised as $\eta_i(t) = \{x(0)\}$;

feedback: The feedback structure can be formalised as $\eta_i(t) = \{x(t)\}$.

---

[17]Throughout, x, u, $g(\ )$ and $f_i(.)$ are assumed to be vectors.
[18]In the engineering literature, the objective function is normally a cost-function that is supposed to be minimised. In the economics profession, it is more standard to maximise objectives (welfare, profit, etc.). The analysis in both cases is essentially the same.
[19]The other three are the 'memoryless perfect state pattern', the 'closed loop perfect state pattern' and the 'ε-delayed closed-loop perfect state pattern' (Başar & Olsder, op. cit. p.212).

The open-loop information pattern describes a situation where the only information assumed to be available for action at time t is that of the initial state. Observation at time t $(t > t_0)$, for instance, the current state is not possible. The other extreme is the feedback case where only the current state is known[20].

Given an information structure, the **strategy** for player i can be defined as a (vector-valued) function $\gamma_i(\eta_i(t), t)$. Therefore, the choice of the player's control variable is: $u_i(t) = \gamma_i(\eta_i(t), t)$. The strategy space Γ_i is the collection of all the possible elements γ_i, given the information set $\eta_i(t)$.

With the definitions above, the game can be solved once an adequate **equilibrium concept** is defined. Here, attention is confined to the Nash and the Pareto outcomes. Other equilibrium concepts, such as the Stackelberg equilibrium will not be considered, as they not be used this chapter.

Definition :
A **Nash solution** is a pair of strategies (γ_1^*, γ_2^*) such that:

$$J_1(\gamma_1^*(t), \gamma_2^*(t)) \geq J_1(\gamma_1(t), \gamma_2^*(t)) \quad \forall\ \gamma_1\ \varepsilon\ \Gamma_1$$
$$J_2(\gamma_1^*(t), \gamma_2^*(t)) \geq J_2(\gamma_1^*(t), \gamma_2(t)) \quad \forall\ \gamma_2\ \varepsilon\ \Gamma_2 \qquad \Box$$

This definition of a Nash equilibrium is a straightforward extension of the Nash concept in static games, with the fixed point (if the equilibrium is unique) being the pair of strategies that are a best response to each other.

The **Pareto solution** (or **cooperative solution**) with weight α can be defined as the optimal pair of strategies (γ_1^*, γ_2^*) for the convex combination of the two objective functionals:

$$\alpha J_1 + (1 - \alpha)J_2$$

This means that, under certain assumptions, the set of Pareto efficient solutions can be found by solving the equation above for all α, $0 \leq \alpha \leq 1$[21]. Typically, however, not all α's give pay-offs that make both players better off. In static games, the set of joint optimisation solutions is therefore often restricted to those weights for which both players are better off (as compared with Nash solution).

In differential games, however, individual rationality may break down along the trajectory. Assume that for a specific α, cooperation is individually rational

---

[20]For comparison, the 'closed-loop' information pattern can be described as $\eta_i(t) = \{x(s), 0 \leq s \leq t\}$. This pattern implies that the information available is that of both open-loop and feedback with additionally perfect recall of all the intermediate values of the state variables.

[21]The notion of Pareto optimality coincides with the concept of group rationality.

at the initial value of the state variables for both players. It is possible that at a certain point along the trajectory, cooperation ceases to be rational for one of the players. Without commitment, this player has an incentive to renege. This time-inconsistency issue will be considered with at the end of the chapter.

Another question is whether the weight α should be fixed over time. In fact, this is not necessarily the case, though there does not seem to be a straightforward way to deal with a time-dependent weight, $\alpha(t)$. Yet another point is how to define α. In the literature, normally either the Nash bargaining solution or the Raiffa-Kalai-Smorodinski solution (cf. Friedman, 1986) are considered, although it would be much more satisfying to actually model the underlying bargaining process. This issue will not be dealt with.

Both the Nash and the Pareto solutions were defined without specifying the type of information structure. Typically, there will be different equilibria for each type of information pattern. For the cooperative solution in a deterministic setting, however, the open-loop and feedback information pattern give the same solution[22]. The Nash equilibrium, on the other hand, differs with the information structure given. For each two of them described above, the Nash equilibrium will be defined by the following theorems (see Başar & Olsder (1982) and Hanig (1986)):

Theorem   (Proof: Başar & Olsder, 1982, p. 279)
If $\{\gamma_1^\star(x_0, t) = u_1^\star(t), \ \gamma_2^\star(x_0, t) = u_2^\star(t)\}$ is an **open-loop Nash equilibrium**, and $\{x^\star(t), 0 \le t \le T\}$ is the corresponding trajectory,
then there exists a costate function $\psi_i^\star(t)$, such that for $i = 1, 2; \ j \ne i$:

$$
\begin{aligned}
\dot{x}^\star(t) &= g(x^\star(t), u_1^\star(t), u_2^\star(t), t) & \text{with } x^\star(0) = x_0 \\
u_i^\star(t) &= \arg \max \mathcal{H}_i(x^\star(t), u_i^\star(t), u_j^\star(t), \psi_i^\star(t), t) & \text{with } u_i \in \Gamma_i \\
\dot{\psi}_i^\star(t) &= -\tfrac{\partial \mathcal{H}_i}{\partial x}(x^\star(t), u_1^\star(t), u_2^\star(t), \psi_i^\star(t), t) \\
\psi_i^\star(T) &= -\tfrac{\partial Q_i}{\partial x}(x^\star(T))
\end{aligned}
$$

where the Hamiltonian $\mathcal{H}_i$ for player i is defined as:

$$
\mathcal{H}_i(x, u_1, u_2, \psi_i, t) = f_i(x, u_1, u_2, t) + \psi_i g(x, u_1, u_2, t) \qquad\qquad \square
$$

The open-loop information structure used here means that both players optimise at time 0 and do not observe the state of the game from that moment on.

---

[22]This is so, as the Pareto solution reduces any multi-player game into a one-player problem. In such cases there is by definition no additional information as time passes. Therefore Pontryagin's maximum principle and Bellman's optimality conditions coincide.

Hence, they cannot react to any deviations from the optimal path. Another way of describing this is that both players have perfect precommitment with respect to their initial strategies.

As an alternative to the open-loop Nash case with only initial state information, the feedback Nash equilibrium is defined next with current state information (and perfect recall of the initial state).

Theorem (Proof: Başar & Olsder, 1982, p. 285)

If $\{\gamma_1^*(x(t), x_0, t) = u_1^*(t), \ \gamma_2^*(x(t), x_0, t) = u_2^*(t)\}$ is a **feedback Nash equilibrium**[23] such that $\gamma_i(x(t), x_0, t)$ is continuously differentiable, and $\{x^*(t), 0 \le t \le T\}$ is the corresponding trajectory, then there exists a (vector-valued) costate function $\psi_i^*(t)$, such that for $i = 1, 2;\ j \neq i$:

$$
\begin{aligned}
\dot{x}^*(t) &= g(x^*(t), u_1^*(t), u_2^*(t), t) && \text{with } x^*(0) = x_0 \\
u_i^*(t) &= \arg\max \mathcal{H}_i(x^*(t), u_i^*(t), u_j^*(t), \psi_i^*(t), t) && \text{for } u_i \epsilon \Gamma_i \\
\dot{\psi}_i^*(t) &= -\frac{\partial \mathcal{H}_i}{\partial x}((x^*(t), u_1^*(t), u_2^*(t), \psi_i^*(t), t) - \cdots \\
&\quad \cdots - \frac{\partial \mathcal{H}_i}{\partial u_j} \frac{\partial u_j^*}{\partial x}(x^*(t), u_1^*(t), u_2^*(t), \psi_i^*(t), t) \\
\psi_i^*(T) &= -\frac{\partial Q_i}{\partial x}(x^*(T))
\end{aligned}
$$

where the Hamiltonian $\mathcal{H}_i$ for player i is defined as before. □

Note that the crucial difference between the open-loop and the feedback formulation is in the additional cross-derivative $\frac{\partial \mathcal{H}_i}{\partial u_j} \frac{\partial u_j^*}{\partial x}$ in the co-state equation. This

---

[23] In order to avoid 'informational non-uniqueness' (see Başar & Olsder, 1982, p. 259-270), the feedback Nash equilibrium is defined as follows:

Definition

The strategies $\gamma_i^* \ \epsilon \ \Gamma_i$ are a **feedback Nash equilibrium** for the differential game with closed-loop information structure if there exist functions $V_i : \Re^n x[0, T] \to \Re$ which for all permissible values of the state variable(s) x_o satisfy:

$$
\begin{aligned}
V_1(x, t) &\ge J_1\{\gamma_1(x, x_0, t), \gamma_2^*(x, x_0, t)\} \ \forall \ \gamma_1 \ \epsilon \ \Gamma_1 \\
V_2(x, t) &\ge J_2\{\gamma_1^*(x, x_0, t), \gamma_2(x, x_0, t)\} \ \forall \ \gamma_2 \ \epsilon \ \Gamma_2
\end{aligned}
$$

□

Note that the so-called '**value function**' V_i must be optimal irrespective of the initial value of the steady state. Therefore, the optimal policy is not derived from a fixed initial condition but as a given function of the initial conditions. Optimal strategies can now be calculated using the so-called Hamilton/Jacobi/Bellman equation:

$$
\frac{\partial V_i(x, t)}{\partial t} + \max_{u_i(x, t)} \{f_i(x(t), u_1(t), u_2(t), t) + \frac{\partial V_i(x, t)}{\partial x} g(x(t), u_1(t), u_2(t), t)\} = 0 \qquad (4.1)
$$

term describes how player i takes the response of player j ($j \neq i$) to changes in the state variable(s) into account. In the open-loop case, the information pattern is such that the evolution of the state variable(s) is not known to the players and therefore, the control variables are not affected by the actual values of the state variables at time t. In the feedback case, the state variables are observed by the players and hence, they may wish to adapt their control variables to changes in x. This reaction by each player is anticipated by the other player and the cross derivative above expresses this effect.

Note that the "*feedback Nash equilibrium strategies will depend only on the time variable and the current value of the state*, but not on memory (including the initial state x_0)" (Başar & Olsder, op. cit. p. 286). This is sometimes referred to as the **Markov restriction**. In fact, we will even limit ourselves further in the rest of the chapter by concentrating on **stationary** games only. These are games where the objective functional $f_i(x(t), u_1(t), u_2(t), t)$ takes the form $e^{-\rho t} f_i(x(t), u_1(t), u_2(t))$ and where furthermore the state equation does not directly depend on time, so that $g(x(t), u_1(t), u_2(t), t)$ reduces to $g(x(t), u_1(t), u_2(t))$. The resulting system of differential equations is said to be **autonomous**. In this case, the feedback Nash equilibrium strategies will only depend on the current value of the state but not on time.

One problem that has not been solved yet is whether **uniqueness**[24] of the equilibrium is guaranteed, even in the linear-quadratic case. It has been proven by Papavassilopoulos & Cruz (1979) that a perfect Nash equilibrium is unique if the time horizon of the game is finite. No uniqueness results appear to be known in the infinite horizon case that is focused on in this chapter (cf. Başar & Olsder, op. cit., p.288). Hanig (1986) hints towards a way to prove uniqueness in the infinite horizon case, when a linear-quadratic value function with linear controls is assumed, but does not formally do so[25]. More problematic is the uniqueness question (in a linear-quadratic framework) using other functional forms of the value function and the controls. In a recent paper, Tsutsui & Mino (1990) showed for a special case (one state variable and infinite time horizon) that multiple equilibria are possible in linear-quadratic models with non-linear strategies[26]. It remains to be seen what

---

[24]This issue of uniqueness is unrelated to the question of 'informational uniquenss', mentioned above.

[25]At the same time, Hanig (1986) is able (using Riccati equations) to calculate for his model all the possible equilibria with linear strategies and to show that within the collection of possible equilibria, there is a unique perfect equilibrium satisfying the transversality conditions.

[26]These strategies are calculated (rather than guessed, as in the linear case) using an auxiliary

the policy implications of the possibility of non-uniqueness are. For other recent discussions on non-uniqueness in differential games, see Clemhout & Wan (1992) and the references therein. In this chapter, it is assumed throughout that uniqueness holds.

An important feature for any intertemporal equilibrium is weak *time-consistency*. This means that restarting the game at a point on the optimal trajectory in any intermediate period t gives a trajectory coinciding with the optimal path of the original game. Both the open-loop and the feedback Nash solutions have this feature. The feedback equilibrium has, however, the additional feature that it remains an equilibrium when the game is restarted from any possible state and not only from the state that is reached along the equilibrium trajectory. This property is referred to as **strong time consistency** or as **subgame perfectness**, which explains why the 'feedback Nash equilibrium' is sometimes named the 'subgame perfect Nash equilibrium'.

In general, the feedback Nash equilibrium is difficult to solve analytically, as dynamic programming techniques have to be used in order to solve the system. An exception is the linear quadratic case, that will be elaborated here.

Linear-Quadratic Differential Games

Linear-quadratic differential games are games characterised by the following simple functional forms: the equations of motion are linear in the state and control variables and the objective functionals are quadratic in the state and control variables. The stationary infinite time horizon version of these games will be analysed here. The objective of player i in such a game is to choose controls u_i so as to maximise[27]:

$$J_i(u_i(t), u_j(t)) = \int_0^\infty [\frac{1}{2}u_i'C_iu_i + \frac{1}{2}x'D_ix + x'E_iu_i + c_i'u_i + d_i'x + h_i]e^{-\rho t}dt \quad (4.2)$$

s.t.

$$\dot{x} = Ax + B_1u_1 + B_2u_2 \qquad x(0) = x_0 \quad (4.3)$$

The dimensions of the matrices A, B_i, C_i, D_i and E_i follow directly, if one assumes that the state variable is a n-dimensional vector, and that player i has m_i control variables. Matrix A and D_i are n x n, C_i is m_i x m_i, whereas B_i and E_i are n x m_i.

---

equation constructed by taking the derivative of the Hamilton-Jacobi-Bellman equation.

[27]This is the infinite time horizon version of the model presented by Hanig (1986) with, additionally, a term expressing the product of x and u. Note that the model by Miller & Salmon (1985) allows for a richer structure (esp. the product u_iu_j is allowed).

These matrices are furthermore assumed to be constant over time[28,29]. The open-loop Nash solution as well as the Pareto solution of this system are straightforward to calculate, using Pontryagin's maximum principle (cf. Miller & Salmon 1985, p.205). For the open-loop case, the first order conditions of a maximum are:

$$
\begin{aligned}
\frac{\partial \mathcal{H}_i}{\partial x} &= D_i x + E_i u_i + d_i + A' \psi_i &&= -\dot{\psi}_i + \rho \psi_i \\
\frac{\partial \mathcal{H}_i}{\partial \psi_i} &= A x + B_1 u_1 + B_2 u_2 &&= \dot{x} \\
\frac{\partial \mathcal{H}_i}{\partial u_i} &= C_i u_i + E_i' x + c_i + B_i' \psi_i &&= 0
\end{aligned}
\tag{4.4}
$$

For the 'cooperative' case, the necessary conditions are similar and will not be elaborated here[30]. For the feedback Nash equilibrium, the optimal strategies can be calculated using the value function $V_i(x)$ and using the Hamilton/Jacobi/Bellman Equation 4.1 for the (autonomous) linear-quadratic case:

$$
\max_{u_i(x)} \{ \frac{1}{2} u_i' C_i u_i + \frac{1}{2} x' D_i x + x' E_i u_i + c_i' u_i + d_i' x + h_i + \frac{\partial V_i(x)'}{\partial x} [A x \;+\; B_1 u_1 + B_2 u_2] \}
$$
$$
= 0 \tag{4.5}
$$

The optimal strategy is then:

$$
u_i^* = -C_i^{-1}(B_i' V_{ix} + E_i' x + c_i) \tag{4.6}
$$

The derivative of the value function with respect to the state variable, $\frac{\partial V_i(x)}{\partial x}$, is the co-state of the system. For the linear-quadratic case, this costate is a linear function of the state alone:

$$
V_{ix} = \psi_i = S_i x + s_i \tag{4.7}
$$

Differentiating Equation 4.7 with respect to time, eliminating $\dot{x}$ and using Equation 4.3 gives an expression that can be compared with the co-state equation. Equating both expressions for $\dot{\psi}_i$, it follows that, if there exists a solution of the

---

[28]The matrix A is negative definite in the example to be discussed below, so that the maximisation of the objective function is, at least in principle, feasible.

[29]Furthermore, c_i is a $(m_i \times 1)$–vector and d_i is a $(n \times 1)$–vector; h_i is a scalar.

[30]Basically, the partial derivative of the Hamiltonian with respect to x becomes:

$$
\frac{\partial \mathcal{H}}{\partial x} = \sum_{j=1}^{2} D_j x + \sum_{j=1}^{2} E_j u_j + \sum_{j=1}^{2} d_j + A' \psi = -\dot{\psi} + \rho \psi
$$

The other first order conditions stay unchanged, as there are no interaction terms $u_i u_j$ $(i \neq j)$.

form $\psi_i = S_i x + s_i$, then S_i and s_i must satisfy the 'Riccati Equations'[31]:

$$
\begin{aligned}
0_{(nxn)} &= A'S_i + S_i A - \rho S_i - (S_i B_i + E_i)C_i^{-1}(B_i'S_i + E_i') - \cdots \\
&\quad - (S_j B_j + E_j)C_j^{-1}B_j'S_i - S_i B_j C_j^{-1}(B_j'S_j + E_j') + D_i \qquad (4.8)
\end{aligned}
$$

$$
\begin{aligned}
0_{(nx1)} &= A's_i - \rho s_i - (S_i B_i + E_i)C_i^{-1}(B - i's_i + c_i - \cdots \\
&\quad - (S_j B_j + E_j)C_j^{-1}B_j's_i - S_i B_j C_j^{-1}(B_j's_j + c_j) + d_i \qquad (4.9)
\end{aligned}
$$

Note also that substituting Equation 4.7 into Equation 4.6 gives the following optimal strategy for player i:

$$
u_i^* = -C_i^{-1}[B_i'S_i + E_i']x - C_i^{-1}[B_i's_i + c_i] \qquad (4.10)
$$

With the Riccati equations and the optimal strategies defined, the following theorems can be stated without proof:

Theorem   (Proof by Papavassilopoulos & Cruz, 1979)
Let the matrices S_i and the vectors s_i satisfy the **Riccati Equations** given above (Equations 4.8 – 4.9), then the strategies u_i^* given in Equation 4.10 are a perfect Nash equilibrium if the matrix Z defined as:

$$
Z = A - B_1 C_1^{-1}(B_1'S_1 + E_1') - B_2 C_2^{-1}(B_2'S_2 + E_2')
$$

is asymptotically stable.   □

Using Equations 4.8 – 4.9, the coefficients of S_i and s_i can be determined and hence, the controls can be expressed as known functions of the state variables. This means that player i knows how player j will react to changes in the state variables. Therefore, the feedback problem can be solved in a similar manner as the open-loop case with Pontryagin's maximum principle by substituting the expression of u_j from Equation 4.10 in Hamiltonian of player i. For player 1, this amounts to:

$$
\begin{aligned}
\mathcal{H}_1 &= [\tfrac{1}{2}u_1'C_1 u_1 + \tfrac{1}{2}x'D_1 x + x'E_1 u_1 + c_1'u_1 + d_1'x + h_1] + \cdots \\
&\quad + \psi_1'[Ax + B_1 u_1 + B_2\{-C_2^{-1}(B_2'S_2 + E_2')x - C_2^{-1}(B_2's_2 + c_2)\}] \quad (4.11)
\end{aligned}
$$

---

[31]Fudenberg & Tirole (1991, p.524-525) and Hanig (1986, p. 62-63) give the Riccati equations for the case that E_i is zero. Following the procedure described above, the 'extended' Riccati equations follow automatically. Note also that in the autonomous differential equation system under scrutiny, the coefficients of the matrix S_i are constant. The general formulation of the Riccati equations with time dependent coefficients is given by Başar & Olsder (1982).

The corresponding first order conditions of a maximum are[32].

$$
\begin{aligned}
\frac{\partial \mathcal{H}_1}{\partial x} &= D_1 x + E_1 u_1 + d_1 + A' \psi_1 - (S_2' B_2 + E_2) C_2^{-1} B_2' \psi_1 &&= -\dot{\psi}_1 + \rho \psi_1 \\
\frac{\partial \mathcal{H}_1}{\partial \psi_1} &= A x + B_1 u_1 + B_2 u_2 &&= \dot{x} \\
\frac{\partial \mathcal{H}_1}{\partial u_1} &= C_1 u_1 + E_1' x + c_1 + B_1' \psi_1 &&= 0
\end{aligned}
$$

This system can now be solved to obtain the trajectories to the steady state of all the relevant variables.

An equivalent procedure is proposed in Miller & Salmon (1985). This will be explained in Appendix B.1. To illustrate the techniques, an example is worked out in Appendix B.2. This finishes the discussion of the theory of differential games.

4.3   Energy-Related Capital in a Multi-Country Model

In this section the one-country model of Section 3.3.1 is extended to a multi-country framework. This allows us to investigate the gains to international cooperation in the field of the environment vis-à-vis the stalemate situation that might arise in the absence of internalisation of the international spill-over effects of GHG emissions. The crucial characteristic of the model in Section 3.3.1 is that GHG emissions can be directly reduced through lower fossil fuel use and indirectly through investment in energy-related capital. In order to model technological progress in the emission reduction sector, the production function is specified as (cf. Equation 3.33):

$$
Y = Y\{e(F,T),P\}
$$

In this specification , $e(.)$ is the energy function. Energy use depends on fossil fuels, F, and on emission reduction technologies, T, in a broad sense, including the result of investment in energy efficiency and renewable resource capital. Output Y depends both on this energy function and on the concentration of Greenhouse gases, P.

In the multi-country extension of this model[33], the equation of motion of the

---

[32]The additional expressions vis-à-vis the open-loop conditions are printed bold; these form the cross-influence term discussed above. See also Miller & Salmon (1985).

[33]For a full description of the model, see Section 3.3.1.

GHG concentration is affected by the total amount of emissions around the globe[34,35].
Hence:

$$\dot{P} = \gamma \sum_{j=1}^{N} F_j - \delta P \qquad (4.12)$$

Assume that the production function and the energy function are the same for all countries. For country i, the social planner's problem will therefore be[36]:

$$\max_{I_i, F_i} \int_0^\infty e^{-\rho t}[U(C_i)]dt \qquad \rho > 0 \qquad (4.13)$$

s.t.

$$C_i = Y_i\{e(F_i, T_i), P\} - I_i \qquad (4.14)$$

$$\dot{P} = \gamma \sum_{j=1}^{N} F_j - \delta P \qquad (4.15)$$

$$\dot{T}_i = I_i - d_T T_i \qquad (4.16)$$

In Section 3.4.1 of the last chapter, energy technology was divided into a physical capital part (K) and human capital part $(H$ and $H_a)$: $T = T(K, H, H_a)$. Investment in H was allowed to have external effects in augmenting the average stock of human capital H_a. In the current section, where the focus is on the spill-over effects of GHG emissions, it is assumed that T does not have any international externality effects. This may be the case because of a credible system of patents[37].

An interesting paper that focuses instead on the international mobility of the results of R & D investment is by Xepapadeas (1991). He has modeled T to include R & D investment in energy technology that could be used to build-up technology in

---

[34]It is assumed that γ is the same for each country.

[35]In pollution problems where the stock of pollution differs per region, the equation of motion in region k would be:

$$\dot{P}_k = \gamma \sum_{j=1}^{N} a_{jk} F_j - \delta P_k$$

where a_{jk} is a coefficient expressing which part of the polluting emissions of region j will eventually fall down in region k (see Mäler, 1989).

[36]Here, the control variables are taken to be I_i and F_i instead of C and F in the previous chapter. This does not change the analysis as such. The reason for the change is that, as will become clear, the solution with I_i as control variable can be more easily tackled in the linear-quadratic case discussed below.

[37]With $T = T(K, H, \bar{H}_a)$ as defined in Chapter, K and H are taken together for simplicity in this section so that the focus is on the dynamics of aggregate energy-related capital T rather than on its physical (K) and human (H) parts separately.

all the regions of the world[38]. In the international pollution game that Xepapadeas describes, the focus is on whether knowledge from R & D investment will be passed on to the other players or not. This is an explicit decision of the players in the game. Cooperation in this game is defined as the situation in which investment augments the global level of human capital. 'Non-cooperation' is taken to describe the situation where national R & D investment only builds up capital at home. Van der Ploeg & Ligthart (1993) take the opposing position that R & D is not patented in the non-cooperative situation.

Alternatively, in the next chapter countries will be allowed to invest in technology in other countries by way of technology transfer. For the time being, however, spending on technology investment by country i are supposed to accumulate technology in country i alone.

As stressed by Currie et al. (1989) cooperation can take different forms, from information exchange to full joint optimisation of objectives. Here attention will be confined to a description of and a comparison between the solution of the full cooperative (or Pareto) case, the open-loop Nash case and the feedback Nash case[39].

As mentioned above, the feedback equilibrium can, generally, only be calculated explicitly for linear-quadratic models. These are models with quadratic objective functions and linear constraints, so that the first order conditions (including the dynamic equations) are linear. However, it is impossible to convert the model given above in a linear-quadratic form, while keeping the essential non-linear characteristics of the production function, $Y_i\{e(F_i, T_i), P\}$[40]. One way around this problem is to linearise around the steady state for each of the equilibria. For the feedback solution, this is a 'chicken and egg' problem, as the steady state is generally not known for models that are not linear-quadratic[41]. Therefore an alternative approach is taken, where the entire model is put into a linear-quadratic form, without linearising the production function. For this purpose, the production function is

---

[38]In his model, there is no physical capital, so that the technology function could be described as $T = T(\bar{K}, H, H_a)$ in the terminology of Chapter 3.

[39]Van der Ploeg & de Zeeuw (1992) also describe the 'market' case or 'no-tax' case. This solution was analysed in the context of the decentralisation of the one-country model of Section 3.3.1. It will not be analysed in the international framework.

[40]Writing production in a logarithmic form is not a way out as the differential equations then become non-linear.

[41]This problem could possibly be solved using an iterative procedure.

substituted into the social welfare function:

$$U(C_i) = U(Y_i\{e(F_i, T_i), P\} - I_i)$$

For this equation, the following quadratic form is used (leaving out the subscripts i):

$$U[Y(.) - I] = \omega_1 T - \frac{1}{2}\omega_2 T^2 + \omega_3 TF + \omega_4 F - \frac{1}{2}\omega_5 F^2 - \frac{1}{2}\omega_6 P^2 - \omega_7 I - \frac{1}{2}\omega_8 I^2 \quad (4.17)$$

where ω_i is a positive constant. In this way the basic concavity assumptions, elaborated in the previous chapter, are maintained[42]. A drawback of this specification is that the functional form of $Y(.)$ cannot be deduced from the welfare function, This means that consumption (C) cannot be calculated. However, $U(C)$ is known so that welfare evaluations are still possible[43]. Note that the dynamic equations of the model are already linear. Therefore, with Equation 4.17, model 4.13 – 4.16 is linear quadratic. In Appendix B.3 , the model is represented in the matrix formulation given in the last section. This specification will be used in the next subsection for the comparison of the open-loop, the feedback and the Pareto case.

4.3.1   Open-Loop, Feedback and Joint Optimisation

Open-loop Nash Case

In the absence of international cooperation, each country chooses the optimal values for its control variables F_i and I_i, taking as given the actions of the other nations. Besides, each nation acts as though it does not care about the environmental damage brought about in other states. This means that the social planner in each country i solves, in the open-loop Nash case, the System 4.13 – 4.16. The necessary conditions, following Pontryagin's maximum principle, are formulated using the Hamiltonian function, $\mathcal{H}_i$[44]:

---

[42]The concavity assumptions are $U'Y_F > 0$, $U'Y_{FF} < 0$, $U'Y_T > 0$, $U'Y_{TT} < 0$, $U'Y_P < 0$, $U'Y_{PP} < 0$, $U'Y_{FT} > 0$, $U'Y_{TP} > 0$. Note that in the previous chapter, these concavity assumptions held without the multiplication by U'. However, as $U(.)$ is concave itself, this does not influence the qualitative results. Note also that, for simplicity, $U'Y_{FP} = 0$ and $U'Y_{TP} = 0$, whereas they were assumed positive in the previous chapter.

[43]In the previous chapter, a deliberate choice with respect to the modelling of the effects of environmental quality was made in favour of productive effects against the alternative of amenity effects. This lead to a utility function of $U(C)$ and a production function of $Y(e(F,T),P)$. A disadvantage of the linearisation given in Equation 4.17 is that this specification could come from a non-linear model with either $U(C,P)$ and $Y(e(F,T))$ or with $U(C)$ and $Y(e(F,T),P)$.

[44]Again the time index is omitted from the equations.

$$\mathcal{H}_i(I_i, F_i, P, T_1..T_N, \psi_i, \xi_{i1}..\xi_{In}) = U[Y\{e(F_i, T_i), P\} - I_i] + \cdots$$
$$\cdots + \psi_i(\gamma \sum_{j=1}^{N} F_j - \alpha P) + \sum_{j=1}^{N} \xi_{ij}(I_j - d_T T_j) \qquad (4.18)$$

where ψ_i and ξ_{ij} are the shadow prices for country i of the stock of GHGs and of the stock of country j's energy technology.

The first-order conditions for internal solutions are:

$$\frac{\partial \mathcal{H}_i}{\partial I_i} = -U_i' + \xi_{ii} = 0 \qquad (4.19)$$

$$\frac{\partial \mathcal{H}_i}{\partial F_i} = \gamma \psi_i + U_i' Y_{If_i} = 0 \qquad (4.20)$$

$$\frac{\partial \mathcal{H}_i}{\partial P} = U_i' Y_{iP} - \alpha \psi_i \qquad (4.21)$$

$$\frac{\partial \mathcal{H}_i}{\partial T_i} = \xi_{ii}(Y_{It} - d_T) \qquad (4.22)$$

$$\frac{\partial \mathcal{H}_i}{\partial T_j} = \xi_{ij}(0 - d_T) \qquad j \neq i \qquad (4.23)$$

Equation 4.23 gives the time-path of the valuation by player i of changes in the capital stock of player j. In the open-loop case, that is focused upon here, this costate is necessarily zero (i.e. $\xi_{ij} = 0$ for $i \neq j$) at each point in time, as players perfectly precommit themselves to an initially chosen path and therefore, act as if they do not put any valuation on deviations of the trajectory of the capital stock of their opponents. Hence, the relevant equations of motion for player i are:

$$\dot{P} = \gamma \sum_{j=1}^{N} F_j - \delta P \qquad (4.24)$$

$$\dot{T}_i = I_i - d_T T_i \qquad (4.25)$$

$$\dot{\psi}_i = (\rho + \alpha)\psi_i - U_i' Y_{Ip} \qquad (4.26)$$

$$\dot{\xi}_{ii} = (\rho + d_T - Y_{It})\xi_{ii} \qquad (4.27)$$

In Appendix B.4, the differential equations and the first order conditions as well as the steady states are given for the linear-quadratic specification of the model.

The meaning of the open-loop solution concept used up till now is that countries do not condition their strategies on observations of the concentration level of pollutants over time and are committed to stick to their initially chosen policies forever. This assumption is perhaps not very realistic, although there are two arguments why open-loop solutions may be satisfactory. Firstly, if investment in emission reduction technology is taken into account, the operational costs of using energy efficient installations will be very small compared with the investment costs and hence there can be a fair amount of commitment. Secondly, it is very difficult to monitor most global environmental problems. Hence, countries must have good faith in each other in order to cooperate internationally in the first place anyhow. On the other hand, one might argue that if the two points discussed here are valid, they should be modelled explicitly.

Feedback Nash Case

In the feedback case, countries' strategies are a function of the current value of the state variable. This means that countries adapt their strategies to changes of the state variables, even if they are unanticipated. To analyse this, define the following so-called value-function, analogous to the social welfare function of the open-loop case[45]:

$$V_i(P, T_1..T_N, t) = \int_t^\infty e^{-\rho t} U(C_i) \, dt \qquad (4.28)$$

subject to:

$$U(C_i) = U(Y_i\{e(F_i, T_i), P\} - I_i) \qquad (4.29)$$

$$\dot{P} = \gamma \sum_{j=1}^N \tilde{F}_j - \delta P \qquad (4.30)$$

$$\dot{T} = \tilde{I}_i - d_{T_i} T_i \qquad \text{i=1..N} \qquad (4.31)$$

where $\tilde{F}_i$ and $\tilde{I}_i$ are the optimal feedback control functions. This value function represents the cost-to-go, given the optimal control. In general, a solution for this system can only be obtained through dynamic programming techniques. However, for the linear-quadratic models, a direct solution of the optimal controls $I_i(P, T_1..T_N)$ and $F_i(P, T_1..T_N)$ can be found, where the controls are written as a linear function of the state variables:

$$F_i(P, T_1..T_N) = s_{i10}P + s_{i11}T_1 + \cdots + s_{i1N}T_N + s_{i1} \qquad (4.32)$$

$$I_i(P, T_1..T_N) = s_{i20}P + s_{i21}T_1 + \cdots + s_{i2N}T_N + s_{i2} \qquad (4.33)$$

As explained in Section 4.2, the coefficients (s_{ijk} and s_{ij}) can be obtained using either Riccati equations or using an iterative method, described in Miller & Salmon

---

[45]Note that time does not enter the value function nor the controls directly here, as the model is autonomous with infinite time horizon, as elaborated in Section 4.2.

(1985), to calculate the trajectory of the variables in the system. In fact this iterative method has been used for the simulation results presented at the end of this section. The system of coupled Riccati equations is given in Appendix B.5.

Note that the specification of the social welfare function[46] includes, with a positive coefficient ω_3, the product $T_i F_i$. This denotes the fact that in the production function the second derivative Y_{FT} is positive, as in Cobb-Douglas functions with substitution possibilities between F and T.

This means that all the coefficients s_{ik} and s_{ikj} are generally non-zero. Note that with $\omega_3 = 0$, the evolution of the stock of capital would be unrelated that of pollution over time. On the other hand, with $\omega_3 \neq 0$, the production side is influenced by the way the pollution externality is dealt with (cooperation versus non-cooperation). Later on in this section, the values of the reaction coefficients will be given for specific numerical values. This will allow a comparison with the open-loop case.

Joint Optimisation

The inefficient solutions given in the open-loop and feedback Nash cases above, is in contrast with the 'terrestrial Paradise' situation where all the externalities are internalised. In this hypothetical benchmark case, a 'global benevolent dictator' optimises joint welfare of all the countries in the world. This means that environmental damages accumulating anywhere in the world are taken into account in national emission reduction policies. The cooperative solution can be calculated by maximising the weighted sum of the social welfare functions of all the N countries:

$$\max_{I_1..I_N, F_1..F_N} \sum_{i=1}^{N} \alpha_i [\int_0^\infty e^{-\rho t} U(C_i)\, dt] \qquad \rho > 0 \qquad (4.34)$$

subject to the constraints given above in the open-loop case. The weights, α_i, denote the importance attributed by the social planner to welfare in country i. Varying the weights gives the collection of all the jointly optimal points. This curve, the so-called Pareto frontier, is depicted for the two-country case in Figure 4.1[47].

Point N in the graph is the joint optimisation solution with focal point weights

---

[46]$U[Y\{e(T,F),P\} - I] = \omega_1 T - \frac{1}{2}\omega_2 T^2 + \omega_3 TF + \omega_4 F - \frac{1}{2}\omega_5 F^2 - \frac{1}{2}\omega_6 P^2 - \omega_7 I - \frac{1}{2}\omega_8 I^2$

[47]The kink in the Pareto frontier is due to the fact that, at a certain value of the weight α, one of the control variables becomes negative. As I have not been able to model non-negativities properly in the model, the dotted part of the Pareto frontier are rough approximations, calculated by hand.

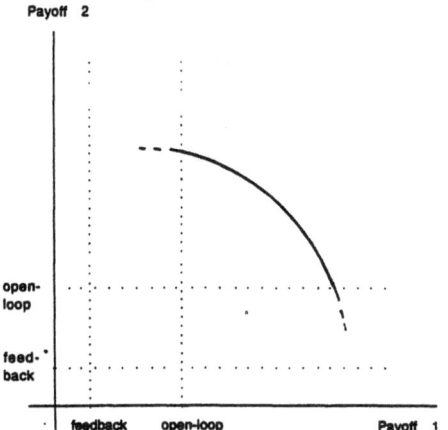

Figure 4.1: *Pareto Frontier*

$\alpha_i = \frac{1}{N}$[48]. The main difference with the non-cooperative open-loop Nash solutions above is that the damages accumulating abroad are incorporated into the analysis, so that marginal global costs instead of marginal national costs are equal to marginal benefits. These so-called Samuelson conditions can most clearly be shown by comparing the equations of motion of the valuation of the GHG stock in the cooperative and the open-loop Nash solution for the focal point case ($\alpha_i = \frac{1}{N}$):

$$
\begin{aligned}
\dot{\psi}_i &= (\rho + \alpha)\psi_i - U'Y_{Ip} & \text{non-cooperative case} \\
\dot{\psi} &= (\rho + \alpha)\psi - \sum_{j=1}^{N} U'Y_{jP} & \text{cooperative case}
\end{aligned}
\qquad (4.35)
$$

The corresponding steady states are $\psi_i^{\infty} = \frac{1}{(\rho+\alpha)}U'Y_P$ for the non-cooperative case and $\psi^{\infty} = \frac{1}{\rho+\alpha}\sum_{i=1}^{N} U'Y_{Ip}$ for the cooperative case. The right-hand side of each expression is the marginal welfare cost of an extra unit of fossil fuel emissions via its detrimental impact on output. Note that in the non-cooperative case, only the impacts at home are measured (i.e. the discounted value of $U'Y_{Ip}$), whereas, in the joint optimisation case, the impacts around the globe of emissions at home are taken into account (i.e. $\sum_{i=j}^{N} U'Y_{Jp}$). This difference also becomes clear by looking at the (P, ψ)–phase plane in Figure 4.2, where the loci for both the open-loop Nash and the Pareto case are compared[49].

---

[48]Ideally, the weights should be determined by real estimations of bargaining power of each country, that can even change over time.

[49]Below, in Figure 4.3, the corresponding time-paths are given for the cooperative case and the two non-cooperative cases. Note that attention is confined to a discussion of the partial phase-diagram of (P, ψ), as all the other state and co-state variables are kept constant. As in the previous chapter, it is assumed that the other variables are at their steady state levels.

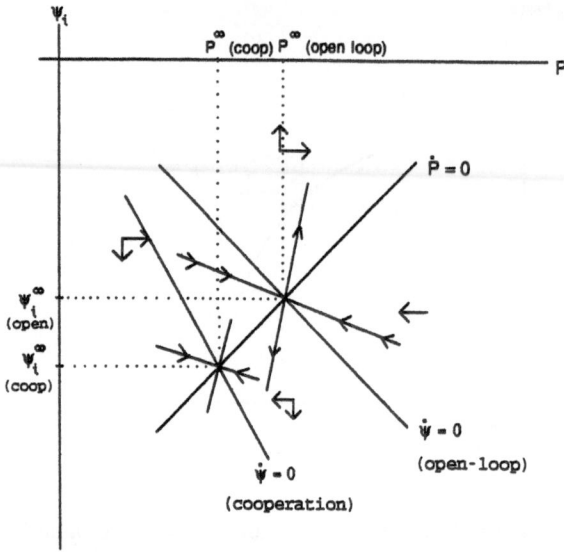

Figure 4.2: *The (P, ψ) Phase Plane for the open-loop Nash case and for the cooperative case*

Note that the steady state level of GHG concentration falls as a result of cooperation. This is due to the fact that the externality of energy use has been internalised in the case of cooperation. In Appendix B.6, it is shown that this is always the case for the linear-quadratic specification of the model.

Note also that, in order to go from an open-loop Nash steady state to a cooperative steady state, the shadow price of the atmospheric GHG concentration will have to jump to a higher absolute value, after which it can fall gradually, in absolute terms, along the new steady state path. This could, in principle, be achieved in a decentralised economy by levying a Pigouvian tax equal to minus the shadow price expressed in terms of C (cf. Section 3.3.1).

Comparison of the Open-loop, Feedback and Pareto Solution

Typically, one would like to compare the steady state results and the dynamic adjustment towards the steady state in the two non-cooperative cases with the results in the joint optimisation case. Van der Ploeg & de Zeeuw (1992) prove for a simple one-state variable model that the steady state level of the GHG concentration is lowest in the cooperative case, and highest in the 'market case'[50]. Less trivially, the steady state concentration for the feedback case is higher than for the corresponding open-loop case. The intuition is that if one country reduced its emissions marginally, this would cause a decrease in the concentration level of GHGs for both countries. In the feedback Nash equilibrium, this country knows that the other country will respond with somewhat lower emission charges, higher consumption and thus higher emissions. In the open-loop case, the countries are precommitted to their initially chosen level of emissions and can thus, by definition, not react to actions of the other country. Hence, the marginal benefits to the environment of an additional unit of emission reduction are lower in the feedback case than they would be in the open-loop Nash equilibrium. Therefore the equilibrium GHG concentration is higher in the feedback Nash Case than in the open-loop Nash equilibrium (Van der Ploeg & de Zeeuw; 1992, p.9)[51].

The argumentation is similar to that of Fershtman & Kamien (1987) in their model on dynamic duopolistic competition with sticky prices[52]. A similar intuition is also used by Reinganum & Stokey (1985) in their oligopoly extraction model of common property natural resources[53].

Given the discussion above, my conjecture is that steady state pollution is higher and welfare lower in the feedback case compared to the open-loop solution. However, this is impossible to prove generally, as the Riccati coefficients cannot be sufficiently simplified. In order to support this conjecture, simulation results are given in Figure 4.3 and Table 4.1 for a set of numerical values for the linearised

---

[50]The 'market case' defined as the situation in which the government does not pursue any environmental policy (as in the decentralised model with $\tau = 0$ in Section 3.3.1)

[51]Note that in a case with uncertainty and with only one actor, the feedback solution is generally better than the open-loop solution. The reason is that the actor can adapt when the circumstances require that. My conjecture that in a situation with both uncertainty and with two actors, the feedback solution can be better or worse (in welfare terms) than the open-loop solution, depending on the specification of the model.

[52]They find in the feedback case that output is higher and profits lower compared to the open-loop solution, stating that it "is not difficult to show that profits are higher at the stationary open-loop equilibrium than at the stationary closed-loop equilibrium. Since the open-loop equilibrium is the case in which each player commits himself to an output path at the outset, and does not condition his output rate on the observed price, it is clear that the players can benefit from such commitments." (Fershtman & Kamien (op. cit. p. 1161))

[53]It is unclear how the issue of possible non-uniqueness, as discussed by Tsutsui & Mino (1990), would affect the discussion on the efficiency of the open-loop solution relative to the feedback outcome. Tsutsui & Mino show that some of the multiple equilibria are more efficient than others, but they do not compare the feedback and open-loop solutions with each other.

version of the model[54,55]. In the figure, the trajectories of the state variable P are compared for the three cases. Analogously, in Table 4.1 the steady states are

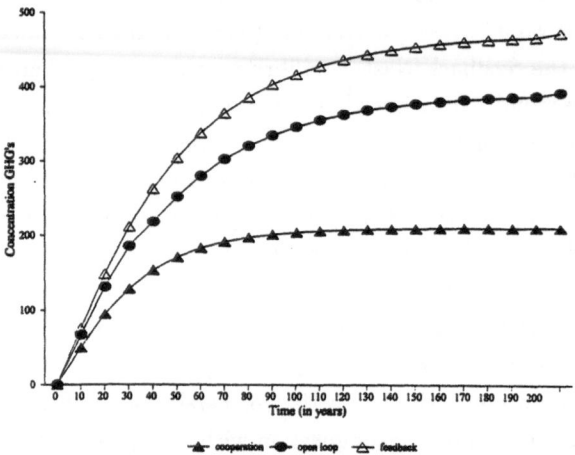

Figure 4.3: *Trajectories in a two country case for the atmospheric concentration of Greenhouse gases (f stands for 'feedback', o for 'open loop' and c for 'cooperation')*

given, showing quite remarkable gains from cooperation [56]. Note that the difference between the feedback and the open-loop solutions are also quite considerable for P and F and for the present value of social welfare. This supports the claim of Van der Ploeg & de Zeeuw (1992) that the benefits of cooperation are considerably higher if the feedback Nash solution is used as a benchmark instead of the open-loop solution. For instance, in welfare terms, a shift from the open-loop solution

---

[54]The values of the coefficients of the social welfare function and the equations of motion are as follows: $w_1 = 2$, $w_2 = 0.01$, $w_3 = 0.05$, $w_4 = 10$, $w_5 = 1$, $w_6 = 0.002$, $w_7 = 5$, $w_8 = 1$, $\alpha = 0.005$, $\gamma = 0.5$, $d_T = 0.1$. Note that the values of α, γ and d_T are the same as in Chapter 3.

[55]The open-loop and the cooperative solution were calculated with the computer-package POREM, developed by Markink & van der Ploeg (1991). This programme gives both the trajectories of the variables, as well as the present value of social welfare. This present value is calculated as the discounted value of the squared deviations from the desired values. The feedback solution was calculated using MARK-2. This is a version of SADDLEPOINT, adapted by Mark Salmon to allow for an iterative calculation of the feedback solution (see Miller & Salmon, 1985). SADDLEPOINT is a FORTRAN model developed by Austin & Buiter (1982). In order to calculate the present value of welfare in the feedback case, the feedback solution was substituted back into POREM. The initial value for P is 0.

[56]The values of the state variables and the controls are augmented with a factor 3, in order to make comparison with the corresponding values of Chapter 3 in quantitative terms more easy.

to the cooperative solution increases welfare with roughly 80 %, whereas starting from the feedback solution increases welfare more than twelvefold.

The time-path for T is similar and will not be shown here. Note that, for the linear-quadratic model here, there is no overshooting, unlike in the non-linear model of Chapter 3.

solution	P^∞	T_i^∞	F_i^∞	I_i^∞	$PV[U(.)]$
cooperation	629.98	182.91	3.15	18.29	525.33
open-loop	1174.86	188.85	5.87	18.89	290.60
feedback	1414.74	193.74	7.07	19.37	42.45

Table 4.1: P^∞, T^∞, C^∞ and F^∞ for a variety of parameter values; case 1 is the benchmark case

Note that energy-related capital is lower in the cooperative case than in the non-cooperative case, as is proven in Appendix B.6. This might be surprising as cooperation means a shift away from fossil fuel towards technology in order to keep production intact while lowering fuel use. On the other hand, due to the lower global level of pollution, a lower quantity of gross production (and therefore less input) is needed to get the same amount of net output $Y\{e(T,F),P\}$. Apparently, this latter 'level' effect is stronger than the 'substitution' effect.

The changes in I_i^∞ and T_i^∞ are only marginal. The magnitude of the change depends crucially on the coefficient w_3 of the product $F_i T_i$ in the social welfare function. If w_3 is put to zero, there will be no effects of cooperation on I_i^∞ and T_i^∞, as is shown in Appendix B.6.

This discussion ends the comparison between the cooperative solution and the Nash solutions. The problem of how to achieve Pareto efficient solutions and how to sustain cooperation have, however, not been dealt with up to now. This is the subject matter of the next section.

4.4   Self-Enforcing International Cooperation

In the previous section, the so-called Pareto Frontier was shown for a numerical specification of the Greenhouse Effect model under scrutiny (see Figure 4.1). The frontier shows the pay-offs of cooperation for both players. Also shown is social welfare in the non-cooperative open-loop and the feedback solution. The difference between the Pareto-optimal pay-offs and the non-cooperative pay-offs denote the gains from cooperation.

In practice, these welfare gains are often not cashed due to, among other things, problems of monitoring compliance and of enforcing treaties. In this section, we focus on how to avoid these problems and how to preserve cooperation theoretically. This will touch on the practical issues mentioned and, hopefully, will deepen the understanding of these issues.

As is well known from the theory of infinitely repeated games, it is possible to sustain individually rational cooperative outcomes as a non-cooperative equilibrium if players adopt trigger strategies and are not too impatient. These trigger strategies imply that, upon deviation from cooperative behaviour by one player, the other players respond with appropriately defined unrelenting punishments. In a repeated game framework, this punishment is typically the one-shot Nash strategy (if unique). Such a punishment only bites as long as cooperation is individually rational (or dominant in the Pareto sense). Crucial is that outside enforcement is not needed: cooperative outcomes are attained in non-cooperative settings. It seems straightforward to extend this idea of the Folk theorem to more sophisticated intertemporal settings. In differential games, for instance, Nash feedback strategies could be adopted as unrelenting punishments to sustain cooperation over time.

However, since "the state evolves over time as a function of the players' controls and the state itself, it may happen that cooperation breaks down in the course of the game when a previously sustainable policy becomes unsustainable or that cooperation emerges when the state reaches a critical level. Consequently, for a cooperative solution to be sustainable as a subgame perfect equilibrium during the whole course of the game, it should give the players higher pay-offs than what they obtain for all (time, state)-pairs under the feedback equilibrium which is regarded as a security level representing the absence of cooperation" (Kaitala & Pohjola, 1990, p.423). This is the issue of time-inconsistency discussed in Section 4.2. In stationary differential games (or pure Markov games) complications arise because Pareto dominance of the cooperative policy depends on the value of the state variable at each instant in time and this value changes typically along the trajectory. There is another complication in defining cooperation in differential games vis-à-vis repeated games. In the latter, trigger strategies lead to punishments in the period immediately following the defection. However, in differential games, as in any continuous time game, this means that reneging could last only an infinitesimal period of time. In order to avoid this problem, one could divide the time horizon into intervals of arbitrary length. This is the idea of δ-strategies (See, A. Friedman, 1971; Tolwinski, Haurie & Leitmann, 1986).

These two complications will be elaborated here to define a 'Folk Theorem' for stationary differential games. Attention is restricted to trigger strategies that sustain cooperation on the Pareto frontier[57]. For this reason, the pay-offs in the Pareto and the non-cooperative case will be defined[58].

Pareto Optimal pay-offs Consider Pareto optimal solutions of an N-player stationary game[59] with weights α[60] and with initial state values $x(t_0) = x_0$. The pay-off of player i at t_0 is defined as: $V_i^c(\alpha, t_0, x(t_0), u_\alpha^\star(.))$. For the Greenhouse game under scrutiny in this chapter, given in Equations 4.13 – 4.16, the vectors of initial states and of Pareto optimal controls are:

$$x_0 = [P_0, T_{i0}, T_{-i0}] \tag{4.36}$$

$$u_\alpha^\star(.) = [F_{i,\alpha}^\star(.), F_{-i,\alpha}^\star(.), I_{i,\alpha}^\star(.), I_{-i,\alpha}^\star(.)] \tag{4.37}$$

Note that $u_\alpha^\star(.)$ depends only on the level of $x(t)$ in stationary differential games so that time does not directly affect the optimal controls[61]. The Pareto optimal continuation pay-off at time t is $V_i^c(\alpha, t, x(t), u_\alpha^\star(.))$. This continuation pay-off represents the 'benefits-to-go' for player i at time t. Observe also that the optimal trajectory $x(.)$ differs for different weights on the Pareto Frontier[62].

Non-Cooperative Feedback Nash pay-offs Define the pay-off for country i in the non-cooperative feedback Nash solution at t_0 as: $V_i^f(t_0, x(t_0), u^\star(.))$. For the

---

[57]A general 'Folk Theorem' for differential games and other types of of dynamic games does not exist. Recent attempts include Lockwood (1990) and Dutta (1991).

[58]The strategy space and the state space are assumed to be finite.

[59]In order to save notation, the N players are denoted as player 'i' and player(s) '$-i$'.

[60]player i has weight α and player(s) $-i$ have weight $(1 - \alpha)$.

[61]It is assumed that the trajectory is unique (see Section 4.2).

[62]For convenience, an indicating subscript α for $x(.)$ is suppressed.

games in the last section, the optimal feedback Nash control is the vector:

$$u^\star(.) = [F_i^\star(.), F_{-i}^\star(.), I_i^\star(.), I_{-i}^\star(.)] \tag{4.38}$$

The feedback Nash continuation pay-off at time t is $V_i^f(t, x(t), u^\star(.))$.

With these definitions, the first complication that is faced with differential games, namely that individual rationality can break down along the trajectory given the state-dependence of the optimal controls, can be tackled (the time-inconsistency issue). Kaitala & Pohjola (1990) use the concept that a Pareto optimal policy $u_\alpha^\star(.) = [F_{i,\alpha}^\star(.), F_{-i,\alpha}^\star(.), I_{i,\alpha}^\star(.), I_{-i,\alpha}^\star(.)]$ is **agreeable** at initial state $x(t_0)$ if its continuation at $x(t)$ dominates the continuation at $x(t)$ of the feedback equilibrium $u^\star(.) = [F_i^\star(.), F_{-i}^\star(.), I_i^\star(.), I_{-i}^\star(.)]$. for all possible values of $x(t)$ along the Pareto optimal trajectory of $x(.)$. This can be formalised as follows:

Definition

A Pareto optimal policy $u_\alpha^\star(.) = [F_{i,\alpha}^\star(.), F_{-i,\alpha}^\star(.), I_{i,\alpha}^\star(.) I_{-i,\alpha}^\star(.)]$ is **agreeable** at initial state $x(t_0)$ if:

$$V_i^c(\alpha, t, x(t), u_\alpha^\star(.)) \geq V_i^f(t, x(t), u^\star(.))$$

this expression should hold for all i, with strict inequality for at least one of the players and for all possible values of $x(t)$ along the Pareto optimal trajectory of $x(.)$.   □

This definition is general in that it deals with all Pareto optimal controls with different weights α and hence it covers all policies at the Pareto Frontier. Kaitala & Pohjola (1988, 1990) restrict themselves to one specific 'efficient' Pareto optimal solution with redistribution[63,64]. In Tolwinski, Haurie & Leitman (1986), all possible cooperative solutions are considered, also those that are not on the Pareto Frontier. This is in line with the 'Folk Theorem'. However, in general, such solutions are not unique in that different trajectories can lead to the same pay-off pair. Hence, Tolwinski, Haurie & Leitman (1986) could not directly define agreeable policies. In

---

[63]In Kaitala & Pohjola (1988), this specific solution is the efficient harvesting programme with transfers. These payments allow one country to buy out the other from the fishery for the purpose of eliminating the inefficiency caused by the joint access to the resources in which one country in better equipped to fish. In Kaitala & Pohjola (1990), this specific solution is the optimal redistributive taxation and optimal economic growth in a game with 'workers' and 'capitalists'.

[64]See also Haurie & Pohjola (1987). Their model deals with dynamic inefficiency of capitalism with 'workers' and 'capitalists' as players. The model in their paper is simpler in that the optimal control strategies are independent of the state variable.

fact, they defined agreeable trajectories (which they refer to as 'acceptable trajectories') along which Pareto dominance does not break down in the course of the game. Their resulting Folk Theorem deals with 'acceptable' trajectories and corresponding trigger strategy pairs instead of focusing on pay-offs. Here, each specific value of the weights α uniquely[65] defines the optimal trajectory and corresponding optimal policies $u_\alpha^\star(.))$ and optimal pay-offs $V_i^c(\alpha, t_0, x(t_0), u_\alpha^\star(.))$.

This allows for a direct definition of 'agreeable' Pareto optimal policies. With this idea of 'agreeable' policies, the first complication posed by the generalisation of trigger strategies in differential games (that individual rationality must be guaranteed along the whole trajectory) has been solved.

The second complication is closely linked with the continuous time characteristic of differential games. To tackle the infinitesimality problem, so-called δ-strategies are introduced (A. Friedman, 1971).

δ-strategies The idea is to divide the time horizon $[t_0, \infty)$ into subintervals of length δ. Each player adopts a 'piece-wise open-loop' strategy which coincides at each subinterval in time with the Pareto optimal feedback strategy $u_\alpha^\star(.)$ (Tolwinski, Haurie & Leitman (1986)). Hence within each subinterval τ, players follow open-loop strategies $u_{\tau,\alpha}^\star(.))$, starting at sample time t_τ[66]. These strategies cannot be adjusted before sample time $t_{\tau+1}$. Assume $t_0 = 0$ and take as time intervals $[\tau\delta, (\tau+1)\delta)$ where $\delta = 1, 2, ..$ (see Kaitala & Pohjola (1990)). The (vector-valued) state variable at sample time t_τ is $x(\tau\delta)$.

Therefore, trigger strategies can be introduced such that the piece-wise open-loop strategy coinciding with the Pareto-optimal feedback is followed, as long as none of the players reneges. However if deviation takes place within subinterval τ (i.e. in the time interval $[\tau\delta, (\tau+1)\delta)$), the players will only discover this at their next check of the state variables at time $t_{\tau+1}$, as $x((\tau+1)\delta)$ does not coincide with the Pareto optimal level of the state variables at $t_{\tau+1}$. As a reaction, the non-cooperative feedback Nash strategy will be followed from $t_{\tau+1}$ on[67]. By allowing the period length δ to approach zero, it is intuitively clear that reneging is unattractive for 'agreeable' Pareto optimal policies. In order to prove this, trigger strategies are

---

[65]Note that uniqueness in the sense that each point on the frontier corresponds to one specific trajectory is, in fact, an assumption (cf. the discussion in Section 4.2 on this uniqueness assumption in the feedback case).

[66]Note that the sample time t_τ is the initial point of time of the time interval $[\tau\delta, (\tau+1)\delta)$.

[67]Trigger strategies where punishment does not last forever can also be considered. Here, however, we restrict ourselves to infinite period punishment.

defined for each value of δ. These strategies are referred to as δ-strategies[68].

Following Kaitala & Pohjola (1990), suppose that each player i has a δ-**strategy** $\hat{u}_\alpha^\delta(.) = \langle \hat{u}_{\tau,\alpha}(.) \rangle_{\tau=0,1,2,..}$ defined as:

$$\hat{u}_{0,\alpha}(.) = u_{0,\alpha}^*(.)$$

$$\hat{u}_{\tau,\alpha}(.) = \begin{cases} u_{\tau,\alpha}^*(.) & \text{if } x(\tau\delta) \text{ equals the Pareto optimal value of } x \text{ at } t_\tau \\ u^*(.) & \text{otherwise} \end{cases}$$

with strategies as defined in Equation 4.37 – 4.38 and with u_α^* taken to be agreeable.

This means that, at initial time t_0, each player will follow the cooperative strategy $u_{0,\alpha}^*(.)$. At the next sample time, t_1, each players checks whether the state reached is the Pareto optimal value of $x(.)$ at that moment. If so, it will pursue the cooperative strategy $u_{1,\alpha}^*(.)$ till the sample time t_2. This goes on as long as no deviation from the jointly optimal trajectory is encountered. Otherwise, each player will follow from the moment of detection of the deviation the non-cooperative feedback strategy $u(.)$.

As a next step, the length of the interval δ between the successive sample times will be decreased. The trigger strategies are now the δ-strategies, where $\delta \to 0$. Formally:

Define **trigger strategies** $\langle \hat{u}_\alpha(.) \rangle$ as infinite sequences of the δ-strategies outlined above:

$$\hat{u}_\alpha(.) = \{\hat{u}_\alpha^{\delta_n}(.), \delta_n \to 0, n = 1, 2, ..\} \tag{4.39}$$

With trigger strategies, thus defined, deviation is not incentive compatible as long as the Pareto optimal policies $u_\alpha^*(.)$ are agreeable, as is stated in the following theorem:

Theorem

The trigger strategies $\langle \hat{u}_\alpha(.) \rangle$ defined in Equation 4.39 above with agreeable strategies u_α^* form a Pareto optimal subgame perfect equilibrium of the differential Greenhouse game defined in Model 4.13– 4.16.

Proof: This follows directly from Tolwinski, Haurie & Leitman (1986). □

Note that every point on the Pareto frontier is assumed to have a unique corresponding trajectory leading to that point. Therefore, the theorem above constitutes a **Folk theorem** for individually rational cooperative solutions on the frontier in the Greenhouse gas model. Hence, by redefining trigger strategies appropriately,

---

[68]See A. Friedman (1971)

the standard 'Folk' argument of guaranteeing cooperation, at least on the frontier, has been maintained in a differential game framework.

It is questionable whether the type of punishment proposed here is realistic. Would the rest of the world deliberately increase fossil fuel use in order to punish a country failing to comply? In practice, often different forms of punishment take place, as in the US Fisherman's Protective Act (Barrett-OECD; Ch.3, 1991). The US allows itself in this Act to punish countries that violate international fishery agreements with trade sanctions. This could be theoretically modeled as the linkage of a fishery game with a trade game. In Chapter 5, this idea of linkage to punish with other means will be extensively dealt with. In the rest of this chapter, the focus is, instead, on functions of the control variables other than trigger strategies which will likewise be able to sustain cooperation, but which will use less severe punishments to deter strategic deviation from the Pareto optimal path.

4.5 Renegotiation Proofness

The original formulation of the Folk Theorem uses infinite period punishment as trigger strategies to deter cheating. These repercussions are, however, very severe. One could, on the other hand, argue that this is not very important as the punishments will, in equilibrium, not be attained anyhow. Still, refinements of the Folk Theorem have concentrated on finite period punishments after which cooperation could be restored. One problem is that the punishments involved hurt the punishers themselves. To understand this, assume that player 1 reneges on cooperation, thereby gaining a one period profit. Knowing that punishment would hurt his opponent, he/she can offer to go back immediately to their cooperative path, as this would be in both players' interest (Pareto superiority). If player 2 cannot resist the temptation to renegotiate, the trigger strategy is deemed unstable, as now "a player will deviate on the expectation that she can convince her opponent to 'forgive' her since it actually (after the deviation has occurred) is in their best interest to do so" (Van Damme, 1989, p.206).

The concept of **renegotiation-proofness** has recently been introduced precisely to avoid such instability of 'self lacerating' punishments. This concept requires "that, in equilibrium, no continuation pay-off be Pareto dominated by another continuation pay-off" (Van Damme, op. cit., p.207)[69]. To explain this, assume

---

[69]Note, this is renegotiation-proofness in the sense of Farrell & Maskin (1989). This is what

a repeated prisoners' dilemma game with pay-off structure (per period) as given in Table 4.2.

Player 2

		D	C
Player 1	D	$(0^\star, 0^\star)$	$(3, -1)$
	C	$(-1, 3)$	$(2, 2)$

Table 4.2: *Pay-off matrix of Prisoner's Dilemma*

As is clear from the table, joint cooperation would give a pay-off of 2 per player per round. A trigger strategy to sustain cooperation could be to play D forever after the opponent has reneged. Note that the continuation pay-off after (C, D) would be $(0, 0)$ forever, which is Pareto dominated by the continuation pay-off after (C, C), being $(2, 2)$ (see Fudenberg & Tirole, 1991, p.180). Therefore, cooperation with trigger strategy (D, D) is not renegotiation proof in the sense described above.

The following 'tit-for-tat strategy'[70] pair, on the other hand, does describe a renegotiation-proof equilibrium that results in the cooperative outcome at every stage: "Begin in the cooperative phase where both players play C. If a single player i deviates to D, switch to the punishment phase for i. In this phase, player i plays C and the other player plays D. Play remains in this phase until the first time player i plays C, at which point play returns to the cooperative phase."[71] (Fudenberg & Tirole, op. cit., p.180).

Formally, this can be rephrased by defining the following three possible contin-

---

Fudenberg & Tirole (op. cit. p.181) call 'internal Pareto consistency'. They also describe other definitions of renegotiation-proofness, the most famous of which is by Pearce. This will be discussed below.

[70]See Axelrod, 1984. Fudenberg & Tirole (1991) refer to this strategy as the 'penance strategy', defining the 'pure tit-for-tat strategy' differently and claiming that strategy is not WRP for the example given above.

[71]It is assumed that the continuation of (D, D) is also (C, C).

uation plays[72,73]:

$$\Pi = \begin{cases} \pi_0 & \text{always play } (C,C) \\ \pi_1 & \text{play } (C,D) \text{ in first round of continuation, otherwise play } (C,C); \\ \pi_2 & \text{play } (D,C) \text{ in first round of continuation, otherwise play } (C,C); \end{cases}$$

For a simple proof that $\Pi = \{\pi_0, \pi_1, \pi_2\}$ is renegotiation proof, see Appendix B.8.

The concept of renegotiation proofness outlined above is from Farrell & Maskin (1989). They refer to it as **WRP (weak renegotiation proofness)**. This idea is not beyond criticism. For instance, a strategy that has as sole continuation plays (D, D) is WRP. This calls for a justification of the possible continuation strategies. The idea is that the concept of WRP of Farell & Maskin is an 'internal' test: the set of continuation plays are taken for granted (a kind of social norm).

However there are **other definitions** of renegotiation proofness, the most famous of which is by Abreu, Pearce & Stachetti (1989). They allow for some of the continuation strategies to Pareto dominate others - and hence they do not test for 'internal' Pareto consistency; instead, they use an 'external' test. Formally, a continuation strategy profile, Π, is renegotiation proof in the sense of Abreu, Pearce & Stachetti unless there is a continuation pay-off v in the collection of pay-offs of the continuation strategies Π, $C(\Pi)$, and another subgame perfect equilibrium Π' such that all the continuation pay-offs in $C(\Pi')$ Pareto dominate v (see Fudenberg & Tirole, 1991, p.181). Put differently, no equilibrium within the solution set should be Pareto dominated by the worst equilibrium in any other self-contained set of equilibria (Farrell & Maskin, 1989). In this case the strategy that has as sole continuation play (D, D) is clearly not renegotiation proof in the sense of Abreu, Pearce & Stachetti (although it is WRP). This can be easily seen by taking as Π' the continuation triplet $\{\pi_0, \pi_1, \pi_2\}$ of the tit-for-tat strategy described above. All the possible pay-offs of this strategy, $C(\Pi')$, are higher than the continuation pay-off (i.e. $v = 0$) of the continuation play (D, D) for high enough discount rates (see Appendix B.8 for pay-offs, where $C(\Pi') = \mathrm{PV}(\rho; \pi_{j,t})$).

This renegotiation concept by Abreu, Pearce & Stachetti (1989) has as drawback that it is *only* 'externally' consistent. Other, stronger, concepts of renegotiation proofness have been put forward (see Farrell & Maskin (1989) and Fudenberg & Tirole (1991)).

---

[72]See Van Damme, op. cit., p.211.

[73]This is continuation strategy when the last outcome was (C, C) or (D, D), as it was assumed that the continuation of (D, D) is also (C, C)

A particularly sensible alternative is mentioned, though not elaborated, in Fudenberg & Tirole (1991). The idea is to take only those WRP strategies of which the cooperative continuation pay-off is Pareto efficient. This eliminates inefficient strategies such as the one mentioned above with sole continuation play (D, D) from the set of renegotiation proof equilibria.

Cave (1987) has introduced a concept of his own in a discrete time dynamic game with static strategies which he calls renegotiation proof equilibrium. According to Farrell and Maskin (op. cit, p. 354), his "concept of renegotiation proofness is that (i) equilibrium pay-offs should be undominated, relative to all subgame-perfect equilibria; and (ii) punishment continuation equilibria are undominated by other feasible sufficient punishments. It seems, therefore, that he seeks relatively efficient punishments to sustain those outcomes that are sustainable using arbitrary (subgame-perfect) punishments". Cave shows that Pareto optimal outcomes can not be supported by renegotiation proof punishments, but that there are outcomes dominating the feedback equilibrium that can be sustained as equilibrium. In what follows, the suggestion of Fudenberg & Tirole (1991) mentioned above will be followed to take as renegotiation proofness concept the idea to choose only those WRP strategies for which the non-reneging continuation pay-off is Pareto efficient. In differential games, the problem with renegotiation proofness is that since the state evolves over time as a function of the players' controls and the state itself, continuation strategies and pay-offs depend on the state. Therefore, attention is confined to agreeable solutions which lie on the Pareto Frontier. Besides, the focus is on differential games with pure Markov strategies, so that the controls are a function of the state alone[74] This allows definition of renegotiation proofness both in terms of pay-offs and in terms of strategies.

4.5.1   Renegotiation Proofness in Differential Games

Any elaboration of renegotiation proofness in differential games gives rise to the same complications as are encountered with trigger strategies in differential games. These complications are two-fold: (1) strategies are state-dependent and (2) infinitesimality causes logical problems for the idea of reneging. These two issues will be tackled here, as above by introducing 'agreeable strategies' and by dividing the time horizon $[t_0, \infty)$ into subintervals of length δ. Concentration is on **two-player**

---

[74]Uniqueness is assumed throughout (see Section 4.2)

games. The introduction of appropriate notation is crucial. Define:

$_j u_\alpha^\star(.)$ is vector of Pareto optimal controls of player j in normal phase;
$_j^i u_\alpha^\star(.)$ is vector of punishment strategies of player j after cheating by i;

As in the last section, piece-wise open loop strategies coinciding with the feedback strategies will be considered. These are defined as follows:

$$_j u_{\tau,\alpha}^\star(.) = \text{coinciding open loop strategy of } _j u_\alpha^\star(.) \text{ in sub-interval } \tau;$$
$$_j^i u_{\tau,\alpha}^\star(.) = \text{coinciding open loop strategy of } _j^i u_\alpha^\star(.) \text{ in sub-interval } \tau;$$

Likewise, the vector-valued controls for players 1 and 2 of the strategies above are $u_\alpha^\star(.)$ and $^i u_\alpha^\star(.)$ for the Pareto optimal controls and the punishments strategies respectively and $u_{\tau,\alpha}^\star(.)$ and $^i u_{\tau,\alpha}^\star(.)$ for the corresponding open loop strategies. So, for instance, $^i u_{\tau,\alpha}^\star(.) = (^i_1 u_{\tau,\alpha}^\star(.), {}^i_2 u_{\tau,\alpha}^\star(.))$

The idea of renegotiation proof strategies is for each player j to start at time t_0 with the Pareto optimal strategy (i.e. $_j u_{t_0,\alpha}^\star(.)$). At the end of each subinterval τ (i.e. at point $t_{\tau+1}$), it will be checked whether the vector of state variables coincides with their Pareto optimal level at that instant of time. If so, the continuation strategy is obviously $_j u_{\tau+1,\alpha}^\star(x(t_{\tau+1}))$. If cheating by i has been noted, on the other hand, the continuation strategy starts with a punishment $_j^i u_{\tau+i,\alpha}^\star(x(t_{\tau+l}))$ (with $l = 1 \cdots k - 1$) during the $(k-1)$ time periods, followed by $_j u_{\tau+k,\alpha}^\star(x(t_{\tau+k}))$ if player i 'behaved well' during the punishment phase[75].

These continuation strategies are chosen in such a way that, in the punishment phase, the punished player is worse off than when he had not reneged and the punisher is better off than in a situation without reneging. After a 'successful' punishment phase, play returns to the Pareto optimal strategy vector. Note, that in a two player game, there are three continuation strategies for each player: $\{u_{\tau,\alpha}^\star(.),$ $^i u_{\tau,\alpha}^\star(.), {}^j u_{\tau,\alpha}^\star(.)\}$ [76].

For the linear-quadratic version of the Greenhouse Effect model, where the Pareto optimal feedback control functions are linear in the state variables, the strategies for fossil fuel are given as (cf. Equation 4.32 with slightly different notation):

$$_j F_\alpha^\star(.) = s_{j1} - s_{j10} P(.) + s_{j11} T_1(.) + s_{j12} T_2 \qquad (4.40)$$

---

[75]The complication that during this punishment phase, player j could renege, will be assumed away.

[76]Note that the issue of existence of a WRP strategy is not dealt with. Only the conditions under which it might exist are elaborated.

If player i reneges in sub-interval τ, the continuation strategy might be to switch for $(k-1)$ intervals $(l = 1 \cdots k - 1)$ to[77]:

$$_j^i F_\alpha^\star(x(t_{\tau+l})) = s'_{j1} - s'_{j10}P(t_{\tau+l}) + s'_{j11}T_1(t_{\tau+l}) + s'_{j12}T_2(t_{\tau+l}) \qquad (4.41)$$

After a 'successful' punishment during these $(k-1)$ intervals, players go back to:

$$_j F_\alpha^\star(x(t_{\tau+k})) = s_{j1} - s_{j10}P(t_{\tau+k}) + s_{j11}T_1(t_{\tau+k}) + s_{j12}T_2(t_{\tau+k}) \qquad (4.42)$$

In order for these strategies[78] to be renegotiation proof, one would typically expect that, among other things, for the punisher $s_{j11} > s'_{j11}$, whereas for the 'ex-cheater', $s_{i11} < s'_{i11}$. This means that the punisher can use more fossil fuel than in the cooperative solution and is hence better off whereas the reneger is obliged to use less fossil fuel than in the cooperative phase. Typically, his best response to punishment gives him a worse outcome than cooperation does. By letting the sub-intervals go to zero, the gains from reneging become small and the negative impacts on the state variable(s) in the cheating interval becomes more and more limited.

This idea will now be worked out, extending Farrell & Maskin's definition of WRP in repeated games. Focus is on linear quadratic models, although my conjecture is that this analysis extends easily to a more general framework[79]. For this purpose, the following notation is introduced. Let the strategy $\hat{\pi}_{\tau,\alpha}(.)$ be the continuation strategy at point of time $\tau\delta$. Then the δ-**continuation-strategy profile** $\hat{\pi}_\alpha^\delta = \langle \hat{\pi}_{\tau,\alpha}(.) \rangle_{\tau=0,1,2,..}$ is defined as:

$$
\hat{\pi}_{\tau,\alpha}(.) =
\begin{cases}
\pi_{\tau,\alpha}^\star(.): & \text{play} \quad u_{\vartheta,\alpha}^\star(.) & \text{in each subinterval } \vartheta = \tau, \tau + 1, .. \\
{}^1\pi_{\tau,\alpha}^\star(.): & \text{play} \quad {}^1u_{\vartheta,\alpha}^\star(.) & \text{for } \vartheta = \tau, .., (\tau + k - 1) ; \\
& \text{and} \quad u_{\vartheta,\alpha}^\star(.) & \text{for } \vartheta \geq \tau + k \\
{}^2\pi_{\tau,\alpha}^\star(.): & \text{play} \quad {}^2u_{\vartheta,\alpha}^\star(.) & \text{for } \vartheta = \tau, .., (\tau + k - 1) ; \\
& \text{and} \quad u_{\vartheta,\alpha}^\star(.) & \text{for } \vartheta \geq \tau + k
\end{cases}
$$

$$(4.46)$$

---

[77]As noted before, it may be more realistic to punish with other means. For this issue, see Section 5.3.

[78]The corresponding controls for investment are:

$$_j I_\alpha^\star(.) \quad = \quad s_{j2} - s_{j20}P(.) + s_{j21}T_1(.) + s_{j22}T_2 \qquad (4.43)$$

$$_j^i I_\pi^\star(x(t_{\tau+l})) \quad = \quad s'_{j2} - s'_{j20}P(t_{\tau+l}) + s'_{j21}T_1(t_{\tau+l}) + s'_{j22}T_2(t_{\tau+l}) \qquad (4.44)$$

$$_j I_\alpha^\star(x(t_{\tau+k})) \quad = \quad s_{j2} - s_{j20}P(t_{\tau+k}) + s_{j21}T_1(t_{\tau+k}) + s_{j22}T_2(t_{\tau+k}) \qquad (4.45)$$

[79]The advantage of linear-quadratic models is that the controls are explicit (linear) functions of the state variable(s).

where $\pi_{\vartheta,\alpha}^*(.)$, $^1\pi_{\vartheta,\alpha}^*(.)$ and $^2\pi_{\vartheta,\alpha}^*(.)$ are the open-loop strategies of the controls defined in Equations 4.40 – 4.45.

These δ-strategies are a formalisation of the three possible continuation strategies at time τ described above[80]. Summarising, these are:

1. No cheating occurred in time interval $\tau - 1$; then the cooperative strategy $\pi_{\tau,\alpha}^*(.)$ will be pursued in τ.

2. Player 1 cheated in time interval $\tau - 1$; then in τ, the penance strategy $^1\pi_{\tau,\alpha}^*(.)$ will be pursued; proceeded by a return to cooperation in $\tau + k$.

3. Player 2 cheated in time interval $\tau - 1$; then in τ, the penance strategy $^2\pi_{\tau,\alpha}^*(.)$ will be pursued; proceeded by a return to cooperation in $\tau + k$.

Having presented the δ–strategy profile, the next step is to let the duration of the time intervals go to zero, so as to decrease the impact of cheating. For this reason, define an **infinitesimal-δ-continuation-strategy profile** $\langle \hat{\pi}_\alpha(.) \rangle$ as infinite sequences of the δ-continuation-strategies outlined above:

$$\hat{\pi}_\alpha(.) = \{ \hat{\pi}_\alpha^{\delta_n}(.), \delta_n \to 0, n = 1, 2, .. \}$$

Note that this strategy profile is the renegotiation proof counter-part of the trigger strategy $\langle \hat{u}_\alpha \rangle$ in the previous section. In order to see this, renegotiation proofness for the Greenhouse gas model under consideration is defined as follows:

Definition
A Pareto optimal equilibrium $\hat{\pi}_{\tau,\alpha}(.)$ of the linear-quadratic version of the differential game defined above with controls given in Equations (4.40– 4.45), is **weakly renegotiation proof** if at any single point of time, none of the continuation equilibria $\{ \pi_{\tau,\alpha}^*(.), {}^1\pi_{\tau,\alpha}^*(.) \text{ and } {}^2\pi_{\tau,\alpha}^*(.) \}$ of $\hat{\pi}_{\tau,\alpha}(.)$ strictly Pareto dominates any other of these continuation equilibria. If an equilibrium $\pi_{\tau,\alpha}^*(.)$ is weakly renegotiation proof, the corresponding pay-offs, $V^c(\alpha, t_\tau, x(t_\tau), \pi_{\tau,\alpha}^*(.))$ are also said to be renegotiation proof. $\square$

Note that this definition is more restricted than the one by Farrell & Maskin in a repeated game framework in the sense that they look at all possible outcomes that can be supported by trigger strategies. In this section, though, concentration is on

---

[80]Note that the penance strategies $^j\pi_{\tau,\alpha}^*(.)$ have a subscript α, whereas their counter-parts $u^*(.)$ of the trigger strategies do not carry a subscript. This is because the punishment of $\hat{\pi}_{\tau,\alpha}(.)$ is not just a function of the state variables, as in the trigger case, but depends on α.

the Pareto optimal strategies. This leads me to the following conjecture in analogy
with Farrel & Maskin's theorem of renegotiation proofness in repeated games:

Conjecture

Let the infinitesimal δ-strategy profile $\langle \hat{\pi}_\alpha(.) \rangle$ defined above form a Pareto optimal equilibrium of the linear-quadratic version of the stationary differential game
defined above, with agreeable strategies π_α^*. Then the following holds:

If for all levels of the state variables along the Pareto optimal trajectory, there
are penance strategies $^j u_{\vartheta,\alpha}^*(.)$ for $\vartheta = \tau, .., (\tau + k - 1)$ (upon deviation by j in time
interval $\tau \dot{-} 1$; $j = 1, 2$) with corresponding continuation actions $^j \pi_{\tau+k,\alpha}^*(.)$, such
that:

1. for the punisher, punishing gives a higher pay-off than immediately returning
   to joint cooperation; i.e.

$$V_i^c(\alpha, t_\tau, x(t_\tau), ^j \pi_{\tau,\alpha}^*(.)) \geq V_i^c(\alpha, t_\tau, x(t_\tau), \pi_{\tau,\alpha}^*(.))$$

2. the best response for a potential cheater, j to the δ-continuation-strategy
   profile of i gives i a lower pay-off than cooperation; i.e..

$$\max_{j\pi(.)} V_j(t_\tau, x(t_\tau), \{ _i\hat{\pi}_{\tau,\alpha}(.), _j \pi(.) \}) < V_j^c(\alpha, t_\tau, x(t_\tau), ^j \pi_{\tau,\alpha}^*(.))$$

then the infinitesimal δ-strategy profile $\langle \hat{\pi}_\alpha(.) \rangle$ and the corresponding pay-offs
are weakly renegotiation proof.   □

In my opinion, this follows directly from the definition given above of weakly
renegotiation proofness and from the fact that the strategies are taken to be agreeable and unique[81]. At the same time, it does not seem trivial to prove this and
I have not been able to do so. In order to clarify the conjecture, assume a discrete time version of the linear-quadratic model with the numerical specification
described above where the interaction term between F and T is set to zero (i.e.
$w_3 = 0$)[82]. The exact specification is given in Appendix B.9. Two cases will be considered: one where the level of the state variable is checked once a year (Table 4.3)

---

[81]Note that this Conjecture only states sufficient conditions while Farrell & Maskin also prove
that these conditions are necessary. Besides, the time intervals are here taken to be infinitesimal.
Therefore, inclusion of a 'large enough discount factor' is not needed here, unlike in Farrell &
Maskin's case.

[82]This basically means that capital accumulation is disconnected from the Greenhouse gas
build-up.

and one where inspection takes place every quarter (Table 4.4). This implies that players can adapt their strategies annually in the first case and quarterly in the second case. The idea of the definition above is that more frequent inspection makes deviation less and less attractive, so that at the limit reneging is not incentive compatible. Hence, every joint cooperative action is renegotiation proof with infinitesimal periods between inspections. In order to illustrate the essence of the Conjecture above, the following cases are distinguished in the tables below:

A Both players cooperate all the time;

B Player II reneges followed by a one year repentance period;

C Player II reneges followed by immediate return to cooperation.

Case B means that with quarterly inspection, there is a four-period repentance time, whereas with annual adjustments of the strategies, reneging is followed by a one-period regret time. Player I is assumed to react in this year by playing his/her feedback strategy[83,84].

Cooperation vs. Cheating with Annual Inspection; 3 possible cases								
year	A: Player I and II cooperate			B: Player II reneges and repents			C: Player II reneges but coop. goes on		
time	F_1	F_2	P	F_1	F_2	P	F_1	F_2	P
1	2.57	2.57	100.0	2.57	10.00	100.0	2.57	10.00	100.0
2	2.57	2.57	102.1	5.75	0.00	105.8	2.57	2.57	105.8
3	2.49	2.49	104.8	2.40	2.40	108.1	2.40	2.40	107.8

Table 4.3: *Cooperation vs. Cheating with Annual Inspection: A Comparison between the values of F and P in the linear quadratic Greenhouse model with $w_3 = 0$*

---

[83]Though other strategies could be thought of as well, such as the one in which player I does exactly what player II did in the year before.

[84]Note that in the case of the Greenhouse Effect, monitoring of the concentration of Greenhouse gases does not necessarily give insight in non-compliance, because many aspects of natural Greenhouse forcing are unknown (e.g. effects of volcano eruptions). Another aspect of this 'signalling' problem is that, with more than two countries (Greenhouse Effect), unexpected changes cannot be unconditionally attributed to one specific non-complying country. Therefore, it would make sense to try and inspect emissions of each individual country rather than the total global build-up of Greenhouse gases.

Comparing case B and case C, it is clear that punishing by the punisher gives a higher pay-off than immediate return to joint cooperation[85]. This means that condition 1 of the Conjecture above is fulfilled.

However, by comparing case A and B, it seems profitable for player II to start cheating and to subsequently repent[86].

quarter	A: Player I and II cooperate			B: Player II reneges and repents			C: Player II reneges but coop. goes on		
Cooperation vs. Cheating with Quarterly Inspection; 3 possible cases									
time	F_1	F_2	P	F_1	F_2	P	F_1	F_2	P
1	0.64	0.64	100.0	0.64	2.50	100.0	0.64	2.50	100.0
2	0.64	0.64	100.5	1.44	0.00	101.4	0.64	0.64	101.5
3	0.64	0.64	101.0	1.43	0.00	102.0	0.63	0.63	102.0
4	0.63	0.63	101.5	1.43	0.00	102.6	0.63	0.63	102.5
5	0.63	0.63	102.0	0.62	0.62	103.2	0.62	0.62	103.0
6	0.63	0.63	102.5	0.62	0.62	103.8	0.62	0.62	103.5

Table 4.4: *Cooperation vs. Cheating with Quarterly Inspection: A Comparison between the values of F and P in the linear quadratic Greenhouse model with* $w_3 = 0$

Comparing this result with the results shown in Table 4.4, note that a one year punishment period is now enough to prevent Player II from cheating. Also the trajectory of the state variable is much less affected than in the case above. In order to check Condition 2 of the Conjecture, other cases than A and B have to be compared as well. In fact, there is an infinite amount of such possible cases to compare. However, in each such case, the basic idea is that with the interval between subsequent inspection points going to zero, is becomes irrational for a player to deviate from the Pareto optimal trajectory. This illustration ends the discussion on how to sustain individually rational cooperation.

---

[85]Note that the relevant part of the utility function is $U_i = 10F_i - 0.5F_i^2 - 0.001P^2$. In order to compare the differences in the 'costs to go', the discounted value of utility has to be calculated, which is quite tedious.

[86]See previous footnote

4.6 Summary and Conclusions

In this chapter, the 'tragedy of the commoms' has been illustrated in the context of a transboundary model of the Greenhouse Effect. For this purpose, a brief review of the theory of differential games was given first, with special emphasis on the difference between the open-loop and feedback Nash equilibrium concepts. This theory was then applied to a multi-country version of the model given in the previous chapter with energy-related capital and a stock of atmospheric Greenhouse gases. The conclusions were that cooperation leads to a higher steady state level of capital and welfare and a lower level of pollution. Furthermore, it was argued that the gains from cooperation are greater with the open-loop Nash solution than in the feedback case, because of the benefit that players have from pre-commitment in the former case.

Note that in models with uncertainty and with only one actor, the feedback solution is generally better than the open-loop solution, as the actor can adapt when the circumstances require that. It is therefore my conjecture that in a situation with both uncertainty and with two actors, the feedback solution can be better or worse (in welfare terms) than the open-loop solution, depending on the specification of the model. It will be an interesting issue for future research to spell out the conditions under which the feedback solution is better or worse than the open-loop equilibrium in welfare terms in models with uncertainty.

Next, trigger strategies and renegotiation-proof strategies were introduced for differential games on pollution. As these games have strategies that depend on state variables that evolve over time, it was argued that cooperation might break down along the trajectory. Also, in differential games, cheating would only last an infinitesimal period of time. This problem was circumvented by the introduction of so-called δ–strategies.

The conclusion is that, in principle, cooperation can be sustained both with trigger strategies and renegotiation-proof strategies, as long as cooperation is not harmful to any of the players. In the former case, a Folk Theorem was given for individually rational cooperative solutions on the Pareto frontier in the Greenhouse gas model. In the latter case, a conjecture was given of when strategies are weakly renegotiation-proof, analogous to Farrell & Maskin's formulation in repeated games. This conjecture was illustrated with a simplified Greenhouse model with different inspection periods. This issue of monitoring is of crucial importance in order to sustain cooperation in international environmental (and other) agreements.

Chapter 5

Cooperation in a Second-Best World: Technology Transfers and Issue Linkage

> *"Ora, per ci che io l'amo, non intendo di voler altra vendetta di lui pigliare, se non quale stata l'offesa: egli ha la mia donna avuta, e io intendo d'aver te. Dove tu non vogli, per certo egli converr che io il ci colga; e per ci che io non intendo di lasciare questa ingiuria impunita, io gli far giuoco che n tu n egli sarete mai lieti."*[1]
>
> G. Boccaccio (1349), Decameron.

5.1 Introduction

The last chapter dealt with cooperation in a first-best world. Pareto optimality could be obtained by jointly maximising the objectives of all the nations involved.

Often, the idea that sovereign states could agree upon some kind of incentive compatible 'coercion' rests on the implicit assumption of rough equality of interest, or symmetry among the relevant actors (Parson & Zeckhauser, 1992, p.4). If Russia

---

[1] Edition of Garzanti, 1980, p. 728; English translation (G. H. McWilliam; Penguin Classics, 1972, p. 648): *"Now, because I love him, the only revenge I propose to take is one that exactly matches the offence. He has possessed my wife, and I intend to possess you. If you refuse to cooperate, I shall certainly catch him out sooner or later, and since I have no intention of allowing his offence to go unpunished, I shall deal with him in such a way as to make both of your lives a perpetual misery."*

were, as is alleged, better off in a 'globally warmed' world, why would it have an incentive for joint action to curb emissions of GHGs? And why would Third World countries with massive underdevelopment and poverty agree to spend money on an issue like Climate Change that may be far beyond the scope of their interest. In such asymmetric situations, welfare gains from cooperation could be distributed such that every actor would still be better off thereby sustaining the joint action.

The apparent benefits of cooperation are in stark contrast to the rather poor record of joint international environmental action. Especially the virtual absence of side payments is striking given the large asymmetries of actors involved. This chapter considers that there are incentive constraints to lump-sum (pecuniary) re-distribution, impeding potentially beneficial (first-best) cooperation. More specifi-cally, two ways of re-establishing some form of joint action are elaborated, giving rise to second-best types of cooperation to curb the Greenhouse Effect. These are:

- donating emission-reduction devices in the form of **technology transfers**;

- combining two roughly offsetting problems by means of **issue linkage**.

First, in **Section 2**, technology transfers of abatement services will be mod-eled. More often than not, less developed countries spend little money on emission reduction. At the same time, most countries in the 'North' are high up on their abatement cost curve. With uniformly distributed pollution, like Greenhouse gases, this means that the 'North' could be better off by investing in abatement in the 'South' than at home. This might also be beneficial to the 'South', even if it were not interested in the Greenhouse Effect as such, because it would typically mean less waste of expensive (often imported) fossil fuel.

With cash transfers, there are often incentive problems with regard to the ques-tion of how the money is actually spent. In this chapter, we argue for in-kind technology transfers as a means of avoiding some of these problems. A simple two-country model is developed for this purpose.

Whether in-kind or money transfers are used, a possible lack of commitment from the 'South' to the continuation of current emission-reduction efforts could make both types of transfers vulnerable to strategic behaviour. This issue is worked out for in-kind transfers and conditions are formulated under which, allowing for such transfers of emission-reduction technology leads to higher consumption and a lower Greenhouse gas concentration. Some simulation results are given to illustrate this point.

In **Section 3**, issue-linkage is introduced as a second way of re-establishing some form of second-best cooperation between asymmetric actors. With issue-linkage, we mean the case of negotiating two separate issues at the same time, with the aim of joint settlement. Given the widespread practice of package deals and horse trading, political scientists have analysed linkage extensively. Game theoretic work in this field includes Sebenius (1983), Tollison & Willett (1983), Stein (1980) and McGinnis (1986).

In the environmental literature, Folmer, van Mouche & Ragland (1991) were the first to work out the idea of linkage. They looked specifically at static supergames, where cooperation is sustained by the use of trigger strategies. Other authors, like Bohm (1990) and Carraro & Siniscalco (1991) have also stressed the potential importance of linkage in international environmental settings.

In this chapter, the analysis of Folmer, van Mouche & Ragland (1991) is extended in several ways. First of all, the concept of differentially valued, roughly offsetting issues is defined. The idea is that full cooperation could be sustained in linked games as long as the 'size of the game' and the asymmetries are such that the games are roughly offsetting. This point is proved using trigger strategies as well as renegotiation proof strategies. This is elaborated in discounted, infinitely repeated two-by-two games. Finally, this idea is extended to differential games, where the stock of pollution changes over time.

5.2 International Cooperation and Technology Transfers

The aim of this section is to present forms of second-best cooperation using technology transfers. For this reason, an intertemporal optimisation model, quite different from the ones in the previous chapters, is developed which is suitable for describing such in-kind transfers. Possible equilibria in the absence of technology transfers will be described first. Next, donations of abatement capital are allowed for and the effect of this change will be analysed.

5.2.1 Description of the Model

The core of the model on technology transfers presented here is Dasgupta (1982). Emission reduction technology is taken as a flow variable. In Chapter 4, technology was more adequately described by a stock variable. However, this would make the analysis here rather difficult and therefore Dasgupta's model has been used and adapted for the purposes of this chapter.

In its one-country model form, a social planner is assumed to maximise welfare intertemporally, which depends on consumption and on the stock of Greenhouse gases. Productive output can be allocated between consumption and expenditures on emission-reduction devices. Emission of Greenhouse gases is related to production via an emission-output ratio. This ratio can be influenced by expenditures on emission reductions (e.g. fuel shifts, investments in the production process itself, introduction of renewable resources, etc.). The atmospheric build-up of Greenhouse gases (GHGs) is influenced by new emissions and by assimilation of GHGs by the environment (cf. Dasgupta, 1982)[2]. The critical trade-off is between present expenditures on consumption giving higher welfare now and present expenditures on emission reduction giving lower 'pollution' and hence higher welfare in the future[3]. This can be formalised as follows[4]:

$$\max_{\alpha,\beta} \int_0^\infty e^{-\rho t}[U(C) + V(P)]\,dt \qquad \text{with } U' > 0,\ U'' < 0,\ V' < 0,\ V'' < 0 \quad (5.1)$$

$$C = 1 - \beta Y \qquad\qquad\qquad (5.2)$$

$$\dot{P} = \epsilon(\beta)Y - \alpha P \qquad\qquad\qquad (5.3)$$

Where: $\epsilon(.)$ = the emission-output ratio;

$1 - \beta$ = the percentage of output devoted to consumption;

β = the percentage of output devoted to emission reduction;

It is assumed that the emission-output ratio $\epsilon(\beta)$ is decreasing and convex in β. In the absence of emission-reduction efforts, $\epsilon(0)$ is defined as the emission-output ratio corresponding to the 'cheapest' production technology. The convex shape of the ratio is due to increasing marginal costs of emission reduction. Note that the social welfare function is additively separate in $U(C)$ (the utility of consumption) and $V(P)$ (the disutility of the Greenhouse gas build-up; $V(P) < 0$). Note also that P does not show up in the production function, unlike the case in the previous chapters (cf. Section 2.3.1). It is assumed that $U(.)$ and $V(.)$ are concave and that $U'(0) = \infty$ and $V'(0) = 0$. The latter means that the first unit of 'pollution' does not influence welfare. Note also that there is no explicit capital depreciation

---

[2]This assimilation is taken to be linear. For a discussion on this issue, see Chapter 3 above.

[3]In the terminology of Section 2.3.1, this is referred to as a 'polluting output model'.

[4]The time index will be omitted from of the equations; As before, C is consumption, P is the stock of Greenhouse gases in the atmosphere and Y is output.

This model is now extended to a two-country setting[5], such that total emissions are $[\epsilon(\beta)Y + \epsilon^*(\beta^*)Y^*]$ instead of $\epsilon(\beta)Y$, where the index "*" indicates the foreign country. Next the possibility is created for the two countries to spend part of their abatement expenditures in the other country. This means that emissions abroad depend on the fraction of foreign output spent on emission reduction in that country as well as on the fraction of output of the home country spent on abatement abroad and vice versa. Additionally, it is assumed that expenditures of the emission reduction by one country in the other country are less effective than expenditures for emission reduction at home. The reason is that less than 100% of the transfer is actually spent on the abatement services. The rest may be spent on transportation, the mark-up of salaries of employees sent abroad, the extra costs due to lack of facilities, infra-structure and so on abroad, etc. The index θ denotes which fraction of the transfer is effectively spent on abatement services. The closer θ is to one, the more efficient the transfer is. Hence, total emissions can be expressed as[6]:

$$\epsilon(\beta_h + \beta_h^*\theta^*)Y + \epsilon^*(\beta_f^* + \beta_f\theta)Y^* \tag{5.4}$$

with θ = index for the effectiveness of the transfer; β_h = fraction of home output spent in the home country; β_f = fraction of home output spent in the foreign country; β_h^* = fraction of foreign output spent in the home country (not *at home*); β_f^* = fraction of foreign output spent in the foreign country; The fractions of output β_f (resp. β_h^*) will be referred to as emission-reduction technology transfers. Note that the transfers stand here for the actual donations of the emission-reduction service and not for the transfers of the 'know-how' alone.

The social planner's problem for country i of the two-country extension of Model 5.1 – 5.3 is[7]:

$$\max_{\beta_h,\beta_f} \int_0^\infty e^{-\rho t}[U(C) + V(P)]dt \qquad \rho > 0 \tag{5.5}$$

$$C = (1-\beta)Y \tag{5.6}$$

$$\dot{P} = \epsilon(\beta_h + \beta_h^*\theta^*)Y + \epsilon^*(\beta_f^* + \beta_f\theta)Y^* - \alpha P \tag{5.7}$$

The Model 5.5 – 5.7 will be used in the following two sections for the analysis of both cooperative and non-cooperative solutions.

---

[5]Expansion to a N-country setting is straightforward.

[6]Here it is assumed that output is the same at home as abroad. This can be easily generalised by adding an index of relative size.

[7]In the absence of technology transfers, Model 5.5 – 5.7 is similar to Van der Ploeg & De Zeeuw (1992) and Cesar (1990).

5.2.2   The Absence of Technology Transfers

The model above will be briefly analysed in the absence of technology transfers (i.e. β and $\beta_h^\star$ are zero)[8]. This establishes a benchmark case to which the merits of introducing technology transfers can be related.

Following Van der Ploeg & de Zeeuw (1992), the subsequent equilibrium concepts will be compared: the full cooperative case, the open loop Nash case, the feedback Nash case and the 'market' case (or 'no-tax' case). The first three cases have been discussed in Chapter 4. The 'market' or 'no-tax' equilibrium describes the situation of an economy consisting of atomistic agents without a national social planner. As the disutility of the Greenhouse gas build-up for each agent is infinitely small compared to the utility of each agents's consumption, the agent is assumed to behave as if his/her actions do not influence the build-up of pollution. This means that the shadow value of the atmospheric concentration of Greenhouse gases is zero. Note that in this solution neither the **intranational** nor the **international** spill-over effects are internalised. The 'market' solution is sometimes referred to as the 'no-tax' solution because the corresponding Pigouvian tax on fossil fuel use is zero. It coincides with marketable permits in which an abundance of emission certificates is available so that the corresponding market price is zero.

In order to see how the optimal path of β, the percentage of income spent on abatement activities, varies with the different equilibria, the (P, ψ) phase plane are drawn for the cases that are considered[9]. Next, the steady state welfare results are compared.

The actual shape of the loci in the (P, ψ) phase plane depends crucially on the specific functional form that seems appropriate in the model. This is briefly worked out for the open-loop case. The feedback case is analysed in Appendix C.1. In the open-loop Nash case, the social planner of the home country solves the System 5.1 – 5.2 with differential equation 5.8.

The static first-order condition with shadow price of the atmospheric concentration of Greenhouse gases, ψ, is:

$$H_\beta = U'Y - \psi\epsilon'Y \le 0; \qquad\qquad H_\beta = 0 \text{ if } \beta > 0 \qquad (5.9)$$

---

[8]The corresponding equation of motion of the GHG build-up in the absence of transfers is:

$$\dot{P} = \epsilon(\beta)Y + \epsilon^\star(\beta^\star)Y^\star - \alpha P \qquad (5.8)$$

[9]The feedback case is not drawn as, for the non-linear specification of the model, analytical solutions cannot be attained.

This means that, in the internal solution where a positive amount of money is spent on emission reduction ($\beta > 0$), the marginal utility of spending a fraction of output less on emission reduction equals the costs in terms of an increase in the future level GHG concentration.

The expression 5.9 can be rewritten as:

$$\beta = 0 \quad \text{and} \quad U' - \psi \epsilon' < 0 \quad \text{corner solution}$$
$$\beta > 0 \quad \text{and} \quad U' - \psi \epsilon' = 0 \quad \text{internal solution}$$

Note that the corner-solution $\beta = 1$ is excluded, as $U'(0) = \infty$.

With an internal solution, β can be written as an implicit function of ψ : $\beta(\psi)$. It can be shown that $\beta'(\psi) < 0$, which is in line with the intuition that a higher value of the shadow price of a GHG build-up in absolute terms[10] leads to a higher fraction of output spent on emission reduction. This implicit function $\beta(\psi)$ can be substituted into the equations of motion of the social planner's problem:

$$\dot{\psi} = (\rho + \alpha)\psi - V'(P) \tag{5.10}$$
$$\dot{P} = \epsilon(\beta(\psi))Y + \epsilon^*(\beta^*)Y^* - \alpha P \tag{5.11}$$

In Figure 5.1, the phase diagram with $\dot{P} = 0$ and $\dot{\psi} = 0$ (denoted as the situation with $V'(P^\infty) = \text{HIGH}$) clarifies this result. The intersection of the two curves is the saddlepoint[11]. Note that the restriction $\beta \geq 0$ means that the $\dot{P} = 0$-locus is kinked at the point $(\epsilon(0)Y/\alpha + \epsilon(\beta^*)Y^*/\alpha; U'(Y)/\epsilon'(0))$.

The corner solution ($\beta = 0$) applies for lower values of ψ in absolute terms. This corresponds to a situation in which the marginal damages at the steady state $V'(P^\infty)$ are too low to induce abatement expenditures. This situation also applies to the 'market' equilibrium where for each atomistic agent marginal disutility of pollution is zero by definition.

Figure 5.1 depicts this situation for the following three cases: $V'(P^\infty) = \text{HIGH}$, $V'(P^\infty) = \text{LOW}$ and $V'(P^\infty) = 0$.

In the cooperative case, damages accumulating abroad are incorporated into the analysis, so that marginal *global* costs instead of marginal *national* costs are equal to marginal benefits, as in the non-cooperative case. Comparing both these cases

---

[10]Note that ψ is negative.

[11]The superscript ∞ denotes the equilibrium value of the variable. The determinant of this system is $\nabla = -\alpha(\rho + \alpha) + V''\epsilon'\beta'(\psi)Y < 0$ and hence the steady state is a saddlepoint.

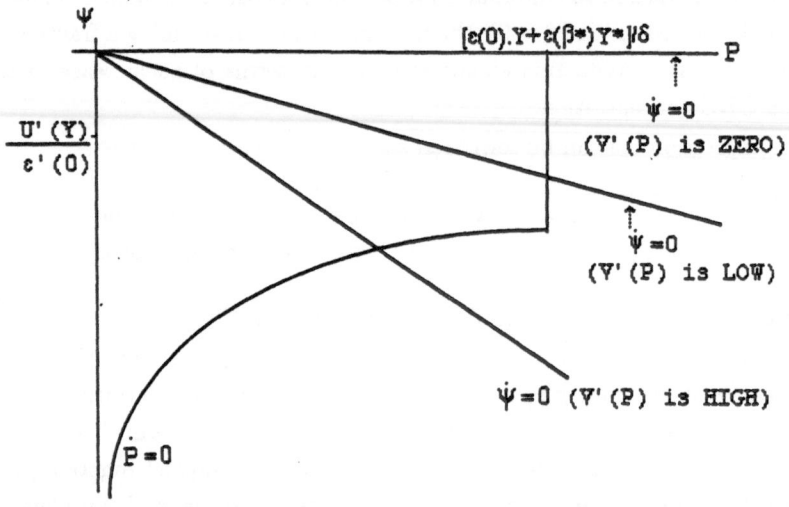

Figure 5.1: *The (P, ψ) Phase Planes for the non-cooperative case without technology transfer ($\beta^\star$ is fixed)*

with the 'market' equilibrium gives the following first order conditions:

$$
\begin{aligned}
U' = \psi_c^\infty \epsilon'(\beta^\infty) &= V'(P^\infty)/(\rho + \alpha) + V'^\star(P^\infty)/(\rho + \alpha) && \text{cooperation} \\
U' = \psi_i^\infty \epsilon'(\beta^\infty) &= V'(P^\infty)/(\rho + \alpha) && \text{non-cooperation} \\
U' = 0 &&& \text{'market' case}
\end{aligned}
$$

The first equation is the famous Samuelson condition for public goods (or 'bad's), guaranteeing a first best social optimum result.

The second difference between the cooperative case and the Nash case is that the $\dot{P} = 0$ locus will shift to the left due to the efficiency gains of allocating abatement in a cost effective way. This is due to the fact that marginal costs of emission reductions for each country are increasing functions of β. In Figure 5.2, the open-loop, the cooperative and the 'market' case are compared.

Note that, in order to go from a open-loop Nash steady state to a cooperative steady state, the shadow price of the atmospheric GHG concentration will have to jump to a higher absolute value, after which it can fall gradually, in absolute terms, along the new steady state path. This could, in principle, be achieved in a decentralised economy by levying a tax equal to the shadow price in absolute terms.

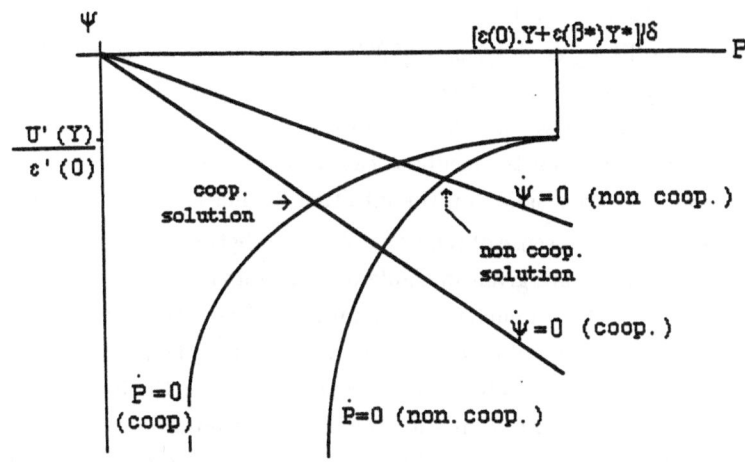

Figure 5.2: *Phase Plane of* (P, ψ) *for the cooperative case, the non-cooperative case and the 'market case' without technology transfer* $(\beta^*$ *is fixed)*

The crucial element in the analysis of the joint global steady state is that the questions of where to abate and who will pay for the abatement are dealt with separately. In practice, both international taxation as well as tradeable emission permits can be used for this means. The decentralisation of the allocation decision and the redistribution decision is crucial for the first best optimum presented above[12]. However elegant in theory, the separation of abatement and payment decisions is not always feasible as damage costs and abatement costs are private information for each country. Besides, countries may not be able to monitor whether the rest of the world is indeed spending their agreed fraction of output on abatement services. In these cases, there are incentive constraints to this joint optimisation, and, therefore, first best solutions may not be attained. Therefore, in-kind technology transfers may be the best available possibility. This will be analysed now.

Typically, one would like to compare the steady state results in the two non-cooperative cases with the results in the joint optimisation case and the 'no-tax'

---

[12]As was shown in the previous chapter, the decision of the initial allocation of pollution certificates in a permit system corresponds to the lump-sum redistribution decision of tax revenues in an emission charge system.

case. As was noted, this is quite difficult, as no analytical solution exists for the feedback case in general.

Van der Ploeg & de Zeeuw (1992) prove for a similar model with a linear-quadratic specification, that the steady state level of GHG concentration is lowest in the cooperative case, and highest in the 'market case'. Less trivially, the steady state concentration for the feedback case is higher than for the corresponding open-loop case. For a discussion on the intuition behind this, see Chapter 4.

In order to get a sense of relative differences between the various solutions, some simulation results are given in Table 5.1 and Figure 5.3 for a *linearised* version of the model[13],[14]. The simulations give an idea of how the relevant variables for the solution concepts described above may differ for specific parameters. In Figure 5.3, the trajectories of the state variable are compared for the four cases. Analogously, in Table 5.1 the welfare results for the different equilibria are given, showing quite remarkable gains from cooperation. Note that the difference between the feedback and the open- loop solutions are relatively small. Table 5.1 also gives the present value of social welfare in the four cases.

solution	stock of GHGs	expenditures in abatement	social welfare $U(C) - V(P) = W(C,P)$		
cooperation	337.38	0.39	15.16	3.95	11.21
open-loop	474.60	0.27	16.15	5.30	10.85
closed-loop	481.94	0.26	16.19	5.39	10.80
'market'	800.00	0.00	16.67	7.62	9.05

Table 5.1: *Simulation results for the steady state of expenditures on emission reduction and of the stock of GHGs, as well as for the present value of welfare*

Note again that welfare is higher in the open-loop case than in the feedback case, although the difference is much smaller with this specification than in Section 4.3.

---

[13]The linearisation is described in Section 5.2.4.

[14]The open-loop, the cooperative and the 'market' solution were calculated with the computer-package POREM, developed by Markink & Van der Ploeg (1991). This programme gives both the trajectories of the variables, as well as the present value of social welfare. This present value is calculated as the discounted value of the squared deviations from the desired values. The feedback solution was calculated using MARK-2 (see Section 4.3. In order to calculate the present value of welfare in the feedback case, the feedback solution was substituted back into POREM. The initial value for P is 0.

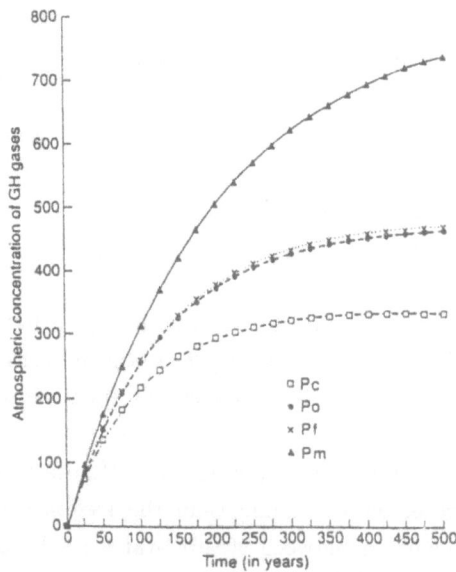

Figure 5.3: *Trajectories in a two country case for the GHG concentration; m stands for 'market', f for 'feedback', o for 'open loop' and c for 'cooperation'*

5.2.3 Cooperation and Non-Cooperation in the Presence of Technology Transfers

In this section the possibility is created for countries to spend part of their output on emission reduction in another country. This will be referred to as the *emission-reduction technology transfer* of one country to another. As noted before, the transfer stands for the actual donation of the energy efficiency and fuel shift services and not for the transfer of the *'know-how'* alone.

This possibility can be incorporated in the model of the previous section by changing the emission-output ratio $\epsilon^*(\beta^*)$ into $\epsilon^*(\beta^* + \beta_f \theta_f)$ as elaborated above with θ denoting the effectiveness of the transfer in the sense of percentage transfer value actually spent on emission reduction[15]. Without loss of generality, it is assumed that transfers go only one way: from the home country (rich) to the foreign country (poor)[16]. In this case the social planner's problem of the rich country is:

---

[15]Note that, as before, it is assumed that $Y^* = Y$ for simplicity.

[16]This means that $\beta_h^* = 0$ and hence, $\beta_f^* = \beta^*$.

$$\max_{\beta_h, \beta_f} \int_0^\infty e^{-\rho t} [U(C) + V(P)] \, dt \qquad \rho > 0 \qquad (5.12)$$

s.t.

$$C = (1 - \beta)Y \qquad\qquad (5.13)$$
$$\dot{P} = \epsilon(\beta_h)Y + \epsilon^*(\beta^* + \beta_f \theta_f)Y^* - \alpha P \qquad (5.14)$$

where $\beta_h + \beta_f = \beta$; hence, β_f denotes the fraction of output that the home country spends in the foreign country; Besides, θ_f is the percentage of the home country's emission-reduction expenditures effectively spent on reductions in the foreign country.

In this section, for the non-cooperative case[17], the focus is on open loop solutions, as they are easier to handle than the feedback solutions. Besides, for the qualitative comparison between steady states in the cases with and without technology transfers, the choice of open-loop versus feedback does not seem to be particularly important. Finally, as the quantitative comparison of the simulation results in the last section shows, the steady state solutions in the feedback and the open-loop case may well be very similar. At the end of this section, however, the feedback case is briefly reconsidered. Assuming internal solutions, the static first order conditions of the non-cooperative case are (cf. Equation 5.9):

$$U_i' = \psi \epsilon_i' = \psi \epsilon_j'^* \theta_{ij} \qquad (5.15)$$

This efficiency condition states that a unit of output spent on either consumption or effective emission reduction should at the margin make the same contribution to welfare. The crucial difference with the (non-cooperative) solution in the absence of technology transfers is that abatement expenditures are now allocated at home and abroad in such a way that marginal costs are equalised in effective terms.

Given the fact that marginal abatement efforts are increasing, this transfer is welfare improving if the emission-reduction efforts of the foreign country are fixed, as is shown is Figure 5.4 for the simplifying case that β^* is fixed.

---

[17]It may be argued that the idea of technology transfers is incompatible with the nature of the non-cooperative solution concept, as transfers require some form of cooperation between sovereign states. However, the point made in this section is that countries, when 'playing' a Nash strategy, are better off in the presence of technology transfers. The Nash strategy as such may include giving each 'player' the right to reduce the other player's emissions.

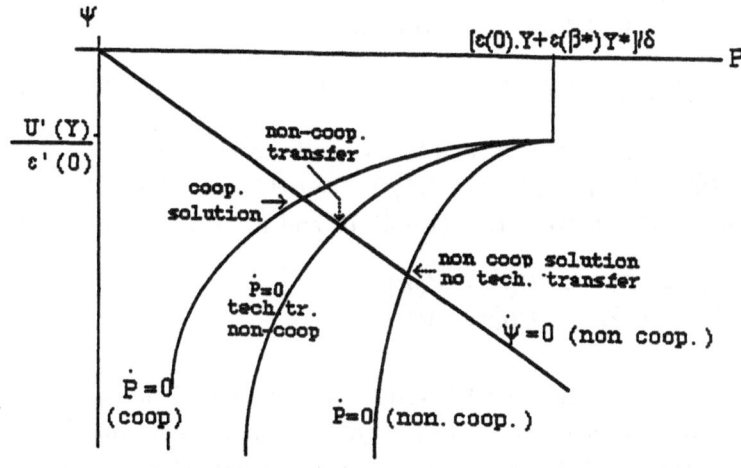

Figure 5.4: *Phase Plane of (P, ψ) for the non-cooperative case with technology transfers (β^* is fixed)*

This idea of welfare gains is formalised in Theorem 5.1.

Theorem 5.1

If at the non-cooperative steady state, marginal emission- reduction costs in effective terms are higher in the home country than in the foreign country (formally, this is: $\epsilon'(\beta^\infty) > \theta_f \epsilon'^*(\beta^{*\infty})$), and if the foreign country can commit itself to fixing emission-reduction efforts at this non-cooperative steady state level ($\beta^{*\infty} = \bar{\beta}^*$),

then, the introduction of emission-reduction technology transfers will lead to a steady state with a lower level of global GHG concentration, a higher consumption level in the home country and hence to a higher level of social welfare in both countries.

Proof:

Define the steady state for the home country with technology transfers as the vector $(\tilde{\beta}^\infty, \tilde{\psi}^\infty, \tilde{P}^\infty)$ and without technology transfers as $(\hat{\beta}^\infty, \hat{\psi}^\infty, \hat{P}^\infty)$.

Then the fact that $\epsilon'(\beta^\infty) > \theta_f \epsilon'^*(\beta^{*\infty})$ together with $\hat{\beta}^{*\infty} = \bar{\beta}^{*\infty}$ implies that:

$$\epsilon(\hat{\beta}^\infty - \tilde{\beta}_f^\infty)Y + \epsilon^*(\hat{\beta}^{*\infty} + \theta_f \tilde{\beta}_f^\infty)Y^* < \epsilon(\hat{\beta}^\infty)Y + \epsilon^*(\hat{\beta}^{*\infty})Y^* = \alpha \hat{P}^\infty$$

Moreover, given $\beta'(\psi) < 0$ and $V'(P) < 0$ and given the steady state equation $\psi^\infty(\rho + \alpha) = V'(P^\infty)$, it is also true that:

$$\epsilon(\hat{\beta}^\infty - \tilde{\beta}_f^\infty)Y + \epsilon^*(\hat{\beta}^{*\infty} + \theta_f\tilde{\beta}_f^\infty)Y^* < \epsilon(\bar{\beta}^\infty - \tilde{\beta}_f^\infty)Y + \epsilon^*(\bar{\beta}^{*\infty} + \theta_f\tilde{\beta}_f^\infty)Y^* < \alpha\hat{P}^\infty$$

This implies that $\hat{\beta}^\infty > \tilde{\beta}^\infty$ and thus $\tilde{C} = (1 - \tilde{\beta}^\infty)Y > \hat{C} = (1 - \hat{\beta}^\infty)Y$. Note also that $\epsilon(\tilde{\beta}^\infty - \tilde{\beta}_f^\infty)Y + \epsilon^*(\bar{\beta}^{*\infty} + \theta_f\tilde{\beta}_f^\infty)Y^* = \alpha\tilde{P}^\infty$. Therefore: $\tilde{P}^\infty < \hat{P}^\infty$.
Given $\tilde{C} = (1 - \tilde{\beta}^\infty)Y > \hat{C} = (1 - \hat{\beta}^\infty)Y$ and $\tilde{P}^\infty < \hat{P}^\infty$, it follows that:

$$
\begin{aligned}
U(\tilde{C}) + V(\tilde{P}) &> U(\hat{C}) + V(\hat{P}) \\
U(\tilde{C}^*) + V(\tilde{P}) &> U(\hat{C}^*) + V(\hat{P})
\end{aligned}
\qquad \square
$$

However, it is, not at all trivial to assume that the foreign country will stick to its pre-transfer steady state level of $\beta^{*\infty}$, as it would imply that the foreign country will stop behaving in a non-cooperative way. One may argue that allowing for technology transfers already implies a kind of cooperation, and hence it is not unreasonable to assume that the foreign country may not play strategically any more. In this case, though, many other steady states may also become reasonable. This point will be worked out later. For the time being, it is assumed that the foreign country will react strategically upon the transfer. In this case it is still possible that both countries are better off. However, the extra constraint is now that the adverse strategic response of the foreign country to the introduction of the technology transfer should not outweigh emission reduction achieved through an optimal allocation of emission-reduction activities. In Figure 5.4, this idea can be interpreted assuming that the reaction of the poor country does not shift the $\dot{P} = 0$ locus all the way back to the initial situation without transfers. This idea will be formalised in Theorem 5.2.

Theorem 5.2

If at the non-cooperative steady state, marginal emission- reduction costs in effective terms are higher in the home country than in the foreign country (that is: $\epsilon'(\beta^\infty) > \theta_f\epsilon'^*(\beta^{*\infty})$), and if the strategic reaction (in terms of emission increases) of the foreign country is not larger than the emission reduction achieved through an optimal allocation of emission-reduction activities (formally: $\epsilon(\hat{\beta}^\infty - \tilde{\beta}_f^\infty)Y + \epsilon^*(\bar{\beta}^{*\infty} + \theta_f\tilde{\beta}_f^\infty)Y^* < \epsilon(\hat{\beta}^\infty)Y + \epsilon^*(\hat{\beta}^{*\infty})Y^*$),

then, the introduction of emission-reduction technology transfers will lead to a steady state with a lower level of global GHG concentration and a higher consumption and welfare level in both countries.

Proof:

Given that $\epsilon(\hat{\beta}^\infty - \tilde{\beta}_f^\infty)Y + \epsilon^*(\tilde{\beta}^{*\infty} + \theta_f \tilde{\beta}_f^\infty)Y^* < \epsilon(\hat{\beta}^\infty)Y + \epsilon^*(\hat{\beta}^{*\infty})Y^* = \alpha \hat{P}^\infty$ holds, and moreover, given $\beta'(\psi) < 0$ and $V'(P) < 0$ and given the steady state equation $\psi^\infty(\rho + \alpha) = V'(P^\infty)$, it is also true that:

$$\epsilon(\hat{\beta}^\infty - \tilde{\beta}_f^\infty)Y + \epsilon^*(\tilde{\beta}^{*\infty} + \theta_f \tilde{\beta}_f^\infty)Y^* < \epsilon(\hat{\beta}^\infty - \tilde{\beta}_f^\infty)Y + \epsilon^*(\tilde{\beta}^{*\infty} + \theta_f \tilde{\beta}_f^\infty)Y* < \alpha \hat{P}^\infty$$

Therefore: $\hat{\beta}^\infty > \tilde{\beta}^\infty$ and thus $\tilde{C} = (1 - \tilde{\beta}^\infty)Y > \hat{C} = (1 - \hat{\beta}^\infty)Y$.   Likewise: $\hat{\beta}^{*\infty} \geq \tilde{\beta}^{*\infty}$ and thus $\tilde{C}^* = (1 - \tilde{\beta}^{*\infty})Y^* \geq \hat{C}^* = (1 - \hat{\beta}^{*\infty})Y^*$ (with equality only if $\hat{\beta}^* = 0$).

Note also that: $\epsilon(\hat{\beta}^\infty - \tilde{\beta}_f^\infty)Y + \epsilon^*(\tilde{\beta}^{*\infty} + \theta_f \tilde{\beta}_f^\infty)Y^* = \alpha \tilde{P}^\infty$ and hence: $\tilde{P}^\infty < \hat{P}^\infty$. Therefore:

$$U(\tilde{C}) + V(\tilde{P}) > U(\hat{C}) + V(\hat{P})$$
$$U(\tilde{C}^*) + V(\tilde{P}) > U(\hat{C}^*) + V(\hat{P}) \hspace{2cm} \square$$

The restriction that $\epsilon(\hat{\beta}^\infty - \tilde{\beta}_f^\infty)Y + \epsilon^*(\tilde{\beta}^{*\infty} + \theta_f \tilde{\beta}_f^\infty)Y^* < \epsilon(\hat{\beta}^\infty)Y + \epsilon^*(\hat{\beta}^{*\infty})Y^*$ need not be problematic in reality. This is especially true in the case of transfers to countries that use a very low percentage of their output for environmental purposes. Put differently, there may be quite some real world examples where the foreign (poor) country is not in the position to 'play the transfer out of the game'. This is especially true if the poor country 'does not have much to play with'.

At this point, it is worthwhile to stress that the feedback solution might have given different results, especially when environmental concerns in both countries are high. This is so, because in the feedback case, the donor country would typically notice cheating by the other country after some time. The donor could, in this case, try to punish the recipient country by stopping the technology transfer. Note, however, that this would generally harm both players (cf. Chapter 4). Therefore, it is not clear beforehand whether it pays for the donor to punish, and this harms the credibility of a possible threat to discontinue the transfer.

Having dealt with the non-cooperative solutions, it is interesting to look into cooperative steady states. Note that, allowing for redistribution, would render the same results as in the full cooperative case described above. Therefore, it is here assumed that redistribution is not allowed. The global social planner solves in this case the following optimal control model:

$$\max_{\beta_h, \beta_f, \beta^*} \int_0^\infty e^{-\rho t}[U(C) + U(C^*) + V(P) + V(P^*)]\, dt \hspace{1cm} \rho > 0 \hspace{1cm} (5.16)$$

s.t.

$$C = (1 - \beta_h - \beta_f)f(K) \tag{5.17}$$

$$C^* = (1 - \beta_h^*)f(K^*) \tag{5.18}$$

$$\dot{P} = \epsilon(\beta_h)f(K) + \epsilon^*(\beta^* + \theta_f\beta_f)f(K^*) - \alpha P \tag{5.19}$$

Note first that, in the absence of additional costs of transfers (i.e. $\theta_f = 1$), this steady state with technology transfers yields exactly the same results as the cooperative outcome with redistribution.

In the case where there are costs of transfers (i.e. $0 \leq \theta_f < 1$), this first best result cannot be obtained. However, as is depicted in Figure 5.5 for realistic values of θ close to one, the second best solution may come reasonably close to the first best result.

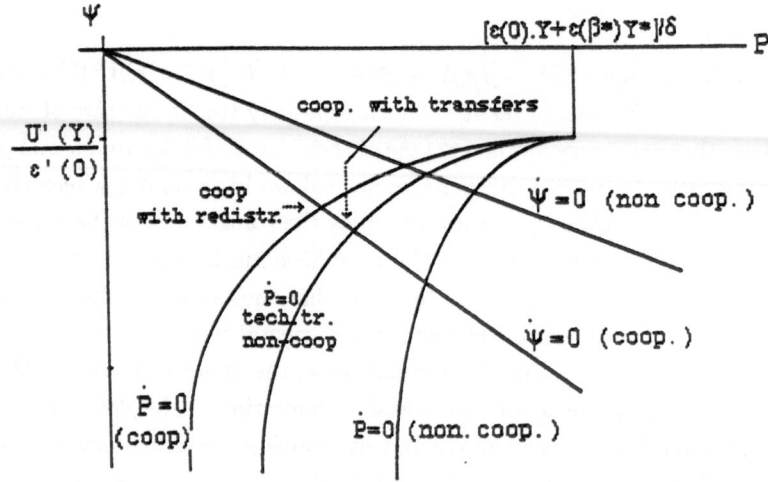

Figure 5.5: *Phase Plane of (P, ψ) for the non-cooperative and the cooperative case with technology transfers (β^* is fixed)*

Here, again, it might be problematic to assume that, in the absence of monitoring, countries will pursue their agreed portion of emission reduction. This leads to a similar analysis as in the non-cooperative case and will not be elaborated here.

At this point it may be worthwhile stressing that technology transfers do not eliminate possibilities of strategic behaviour, though it may limit the scope of such behaviour. Basically in the case of joint global optimisation with lump sum redistribution, countries could be tempted to:

- overestimate the damages of a GHG build-up (cheat with $V(P)$);

- overestimate the abatement cost function (cheat with $\epsilon(\beta)$);

- carry out less emission reduction than agreed upon (cheat with β);

In the case of technology transfers, receiving countries still have the possibility to overestimate $V(P)$, though donor countries now control $\epsilon(\beta)$. Besides, due to the transfers, donor countries will tend to get a better insight in the true emission-reduction expenditures of the receiving country.

In a non-cooperative case, the only remaining incentive problems of the receiving countries are the strategic decrease of expenditures on emission reduction in the light of the transfers. This issue can typically be better worked out in a bargaining framework than in the type of modelling pursued here.

It has been argued that the potential gains from technology transfers are considerable. Implicit in this argument is the assumption that θ, the parameter of the effectiveness of transfers, is close to one. In the simulation results in the next section, for instance, this parameter is set to 0.75. At the same time, the use of technology transfers in development aid programmes has not been an unqualified success. In fact, as has been pointed out[18], it may well be that θ is close to zero with the actual gains of technology transfers being hence substantially lower. The effectiveness of transfers might, in reality, be restricted due to lack of infrastructure, human capital, labour discipline an so forth. Therefore, the actual size of the welfare gains of technology transfers remains uncertain and is an empirical question.

5.2.4 Simulation Results

In order to illustrate the importance of technology transfers in an open-loop non-cooperative setting, a linear-quadratic version of the model has been used to generate some quantitative results[19]. For this reason, the following three cases are distinguished:

---

[18]Michael Hoel pointed this out to me; See also Stewart (1990)

[19]The linear-quadratic version is different in the model presented in Equations 5.5 – 5.7. The reason is that it is impossible to capture the specific convex character of the emission- output ratio in a linear-quadratic game (quadratic objective functional and linear equations of motion). In order to get around this problem, investments in emission-reduction technology, I are introduced. Each country can optimise net social welfare using as controls both Y and I. Y is included both in a

- emission reduction in both countries without technology transfers;

- emission reduction with technology transfers in the absence of precommitment of the receiving country with respect to its own emission reduction policy;

- emission reduction with technology transfers in the presence of precommitment of the receiving country with respect to its own emission reduction policy.

The results are presented in Table 5.2[20]. Two situations are distinguished. In the first case, the countries are identical in every aspect (the benchmark case). This case was presented previously before in Table 5.1. The second situation is characterised by the difference in the perception of the damages due to the accumulation of GHG emissions between the two countries, expressed by the constant c[21]. For the rest, the countries are identical.

As is clear from Table 5.2, the precommitment issue is not important as long as the receiving country does not perceive any disutility from the Greenhouse Effect (no expenditure on emission reduction). This is in line with the comments in Section 5.2.3. In the case in which both countries care about the Greenhouse Effect but one cares more than the other, different results are obtained for the situations 'no transfers', 'transfers with precommitment' and 'transfers without precommitment'. The directions of the changes are in line with the remarks in Section 5.2.3. For instance, compared to the 'no transfer' case, the introduction of transfers without precommitment lead to higher net social welfare for both countries. At the same

---

linear and in a quadratic form in the objective functional whereas I is only included in a quadratic form. Both Y and I are linearly included in the equation of motion. In this way, the character of an emission-output coefficient is mimicked. The model is: $U(C) = 0.5 + Y - 0.25Y^2 - 50I_h^2 - 50I_f^2$, $V(P) = 1000cP^2$ and $P = [0.1Y - I_h] + [0.1Y^* - I^* - \theta I_f] - 0.005.P$. For the calculations the constant θ is taken to be 0.75. The constant c ranges from 0 to 4, depending on the specific version of the model, as shown in Table 5.2. An obvious objection to the specification is that the costs of technology transfers are underestimated, as it is assumed that these costs are independent of the amount of emission reduction already carried out. However, this may not be too severe if it is assumed that relatively low emission-reduction efforts take place in the receiving country (as is done in Table 5.2). Besides, qualitative results will not alter, though quantitative results may be less pronounced.

[20]The open-loop solutions were calculated with the computer-package POREM, developed by Markink & van der Ploeg (1991). This programme gives both the trajectories of the variables, as well as the present value of social welfare. This present value is calculated as the discounted value of the squared deviations from the desired values. Initial values for P are 0.

[21]The constant c denotes the relative importance of $V(P)$; $V(P) = 1000cP^2$.

	envir. concern		stock GHGs	expenditures in emis. red.			social welfare			
solution	c	$c^\star$	P	I_h	I_f	$I^\star$	$U(C)$	$U(C^\star)$	$V(P)$	$V(P^\star)$
benchmark	2	2	474.6	0.27	0.00	0.27	16.15	16.15	5.30	5.30
no transfer	4	0	474.6	0.54	0.00	0.00	14.58	16.67	10.61	0.00
transfers without precomm.	4	0	440.9	0.50	0.38	0.00	14.34	16.67	9.99	0.00
transfers with precomm.	4	0	440.9	0.50	0.38	0.00	14.34	16.67	9.99	0.00
no transfer	3	1	474.6	0.41	0.00	0.14	15.50	16.54	7.96	2.65
transfers without precomm.	3	1	448.9	0.39	0.29	0.13	15.34	16.55	7.61	2.54
transfers with precomm.	3	1	446.1	0.38	0.29	0.14	15.35	16.54	7.59	2.53

Table 5.2: *Simulation results for the steady state of P and I and for the present value of welfare for different values of c*

time, expenditures on emission reduction at home drop for both countries and so does the stock of GHGs. If the receiving country were able to precommit itself to adhere to its 'pre-transfer' level of emission reduction, the welfare gains for the donating county would be more pronounced. This is due to the additional decrease in the steady state value of atmospheric GHG concentration.

This example, illustrating the two theorems presented here, finishes the discussion on technology transfers. In the next section, a rather different form of second best cooperation will be considered: issue linkage.

5.3 Issue Linkage in International Environmental Problems

5.3.1 Introduction

Package deals, horse trading and coercive cooperation are all too familiar practices in international negotiations. The common element of these practices is the use of linkage of (not necessarily related) issues to reach an agreement. However widespread linkage may be in many international negotiations, this chapter argues that in international environmental bargaining situations, linkage has not yet been used as much as it could have been.

Issues are said to be linked (or 'added' or combined) when they are simultaneously discussed for joint settlement[22]. Raiffa (1982) in his seminal book on negotiations stresses the potential of linkage in multi-issue negotiations to achieve joint gains. In his comprehensive paper on linkage, Sebenius (1983) suggests that issue linkage can be used to:

1 Give leverage to a hegemon;

2 Create a zone of possible agreements between parties;

3 Destroy a zone of possible agreements between parties.

A famous example of linkage to give leverage to a hegemon is the linking of US support for the British pound with the discontinuation of Britain's involvement in the Suez crisis. Linkage by the U.S. 'persuaded' Britain to change their military policy in the Middle East (Stein, 1980, p.68). An often-mentioned example of creation of a zone of possible agreements is the linkage of deep seabed mining with maritime issues in the Law of the Sea Conference (Sebenius, 1983). Another example can be found in the London Amendments of the Montreal Protocol where a phase-out of CFCs in developing countries was linked with technology transfers (via the Ozone Trust Fund) (Parson, 1992). An example where zones of possible agreement were destroyed is given by Fisher (1964, p.93): "In August, 1961, a civil aviation agreement between the United States and the Soviet Union was negotiated. The United States might have signed the agreement, treating it as a separate matter. We chose, however, to decline to sign it, and considered the matter related to Berlin"[23].

This section will focus especially on the second type of situations, where a zone

---

[22]Sebenius (1983, p.288).

[23]Fisher (1964, p.93). I found this example in Sebenius (1983, p.200).

of possible agreements (ZoPA) between parties may be created or enhanced through linkage[24]. It seems that **asymmetry between actors and between issues** plays often a crucial role (Raiffa, 1982). First, due to the asymmetry between actors on each issue, a ZoPA may not exist on each issue or alternatively there is no focal point within the ZoPA (Parson & Zeckhauser, 1992). Moreover, asymmetry between issues invites linkage, just as differences in endowments between individuals creates possibilities for mutually beneficial trade.

Sebenius (1983, p. 293) distinguishes between three such cases where a ZoPA can be created by linking issues:

A   creating or enhancing a ZoPA by adding differentially valued, unrelated issues;

B   overcoming distributional impediments to jointly beneficial agreements by adding issues as side payments;

C   adding issues to exploit their dependencies.

Case C is particularly relevant when there are synergy effects between the issues. Tollison & Willett (1979) give a nice example of case B, assuming increasing returns to scale in military high-tech. Hence, it would be optimal to build this high-tech equipment in one country, where the other country would contribute to the project via side-payments. When side-payments are not feasible, two similar projects can be taken together where one country produces one type of high-tech equipment and the other country another type.

In this chapter, I will concentrate on case A, where a ZoPA can be created or enhanced by adding differentially valued, roughly offsetting issues. Specifically, I will look at linkage of two games, rather than linkage of one game with one non-game (e.g. side payments). In game-theoretic terms, this means that the controllable subspace is enlarged and the loss function is altered[25].

As prime example, I will take differences in opinion on the Greenhouse Effect negotiations between different OECD countries. It seems that this issue is differently valued in the US than in the rest of the OECD (ROECD), though this may be changing right now with the new Clinton-Gore administration. A roughly offsetting issue could be international security, where the ROECD seems to be less inclined to invest in their defense than the US. Linkage between these two issues may thus

---

[24]Note that this depends ultimately on the specification of the loss function.

[25]The question could be raised why issues are not linked in the first place if that is better in welfare terms. One possible answer is that unlinked issued are typically discussed at a lower administrative level than linked issues. Combining an environmental with a trade game may then mean that prime minister's offices of two countries negotiate instead of the home and foreign environmental ministries on one issue and their counterpart trade ministries on the other issue.

dramatically alter the negotiation results.

At the same time, linkage is not without drawbacks. Tollison & Willett (1979) argue, for instance, that linkage typically augments the number of effective decision makers, thereby increasing decision costs. Sebenius (1983) states that to add issues is to "run the risk of increasing complexity, causing information or analytic overload or fear of unintended consequences, thereby burdening the negotiation itself with possibly significant costs"[26]. Hence, there seems to be a trade-off between the virtues of linkage and the added complexity. Therefore, sometimes, "issue-by-issue considerations may be necessary to avoid the sheer complexity of combination and partial agreements may provide impetus to overall ones"[27]. In other situations, like the Law of the Sea negotiations, the very fact that the manifold issues were considered together in à single mammoth negotiation, so that a concession on one point could be balanced by a concession in another, made a compromise between the conflicting interests possible[28]. Keohane (1984) stresses the importance of regimes to link issues efficiently. His reasoning is, similar to the arguments given above, that the existence of regimes lowers the transaction costs of taking issues together. "Regimes are relatively efficient institutions, compared to the alternative of having a myriad of unrelated agreements, since their principles, rules, and institutions create linkage among issues that gives actors incentives to reach mutually beneficial agreements. They thrive in situations where states have common as well as conflicting interests on multiple, overlapping issues and where externalities are difficult but not impossible to deal with through bargaining"[29].

Viewing the Greenhouse Effect as a long term security question and hence placing military security and the Greenhouse effect in one regime, may thus enhance cooperation in both issue areas. I strongly believe that this form of linkage is not as far fetched as it may sound, given the intimate relationship between energy policy and the Greenhouse effect. Energy policy, in turn, has been considered somewhat tied to security issues in the Carter era and is still linked to national security in many other Western countries (See also Ullman, 1983).

Hence, specific reciprocity between these two issues will be advocated here, where reciprocity refers to "exchanges of roughly equivalent value in which the actions of each party are contingent on the prior actions in such a way that good

---

[26]Sebenius (1983, p.206).

[27]Sebenius (op. cit. p.298).

[28]Sebenius (op. cit. p.299).

[29]Keohane (1984, p.97).

is returned for good and bad for bad" (Keohane, 1986, p.8).

Above, it was claimed that asymmetry is crucial for **establishing** cooperation. However, it will turn out that asymmetry is also instrumental in **sustaining** cooperation. The reason is that each country can use the possible discontinuation of cooperation in the issue area it is less interested in as a threat against possible failure by the other country to cooperate in the issue area that the first country is most interested in. In this way, cooperation is self-enforcing.

Sustaining cooperation through self-enforcing agreements is especially important in international negotiations, as the anarchic structure of world politics means "that the achievement of cooperation can depend neither on deference to hierarchical authority nor on centralised enforcement. On the contrary, if cooperation is to emerge, whatever produces it must be consistent with the principles of sovereignty and self-help"[30].

In this chapter, it will be argued that full cooperation can be sustained in linked games as long as the 'size of the game' and the asymmetries are such that the games are roughly offsetting. This point will be formalised and proved using trigger strategies as well as renegotiation proof strategies. This will first be elaborated in discounted, infinitely repeated two-by-two games[31]. Next, linkage in differential games is considered. As elaborated in Section 4.5, the advantage of renegotiation proof strategies is that, once a country has failed to cooperate for some period of time, cooperation can be restored after a period of repentance. Tit-for-tat strategies are often renegotiation proof with the additional advantage that they have a minimal punishment. The concentration on two-by-two games raises the question of the extent to which the results can be generalised to more realistic settings. However, as Martin (1992-b) argues, it is important to note that although only two alternatives are typically distinguished — full sanctions and no sanctions — each player "in fact decides on a value bounded by 0 and 1. The two-by-two matrix does not show the full game, only the boundary pay-offs"[32].

In its nature, this chapter comes closest to McGinnis (1986), who focuses on situations where cooperation via linkage can be expected and where the mere repetition of the games over time is not sufficient to sustain cooperation. Folmer, van Mouche & Ragland (1991) argue, in their example, for linkage as an alternative

---

[30]Keohane (1986, p.1).

[31]The requirement of 'infinity' can be replaced by the concept of a finitely repeated games with a probability that the game will end in every period.

[32]Martin, 1992-b, p.17.

for politically unfeasible side-payments, where in each 'linkable' issue, a game is created only by including the transfer payments to this issue[33].

Novel in this chapter are the stress on asymmetry in sustaining cooperation, the formalisation of the idea of 'roughly offsetting' games, the introduction of renegotiation proof equilibria in the field of linkage, and the use of linkage in differential games.

5.3.2   The Analytical Framework for Two-By-Two Games

For the cases that are being discussed, the following assumptions are made:

- there are two countries (players), seen as rational monolithic actors;
- two different issues are being negotiated at the same time;
- each issue is seen as an infinitely repeated game;
- there are only two strategies per issue: cooperate or defect;
- concentration is on pure strategies only.

Hence, each game can be represented in the following pay-off matrix, as given in Table 5.3.

		Player 2	
		D	C
Player 1	D	(n_1, n_2)	(f_1, u_2)
	D	(u_1, f_2)	(m_1, m_2)

Table 5.3: *Pay-off matrix of a general two-by-two game*

In the case of pollution, this matrix represents a two-player game where the players have the choice between abating a fixed amount or not abating at all (C resp. D). For instance, element f_2 is the pay-off to player 2 if he/she does not cooperate, while the opponent does cooperate. I assume that inter-player comparison of pay-offs is possible. The present value of pay-offs for player i with discount rate ρ is defined as[34]:

$$PV_i(\rho) = \rho \sum_{t=0}^{\infty} (1-\rho)^t pv_{i,t} \qquad (5.20)$$

---

[33]This approach is also taken by Stein (1980) in his Suez crisis example, discussed above.

[34]$pv_{i,t}$ is defined as the pay-off in period t of player i.

According to the Folk theorem, the cooperative solution (m_1, m_2) can be sustained as a (subgame perfect Nash) equilibrium using an appropriate threat strategy, as long as for each player, its cooperative pay-off is larger than its non-cooperative equilibrium pay-off and as long as the discount rate is small enough (J. Friedman, 1971). It seems that in a lot of the cases in literature where issue linkage is suggested, the mere repetition over time would be sufficient to guarantee cooperation.

Hence, the focus will be on situations in which the joint cooperative pay-off is larger than the non-cooperative equilibrium pay-off but where the non-cooperative equilibrium pay-off is larger for one player than the cooperative pay-off[35]. In game-theoretic terms, this says that cooperation is maximising joint pay-offs but is not individually rational (not Pareto superior). In those cases, linkage of infinitely repeated games is necessary to reach cooperation (Mc Ginnis, 1986).

To give a numerical example, one can think of abatement costs as being 6 per unit for country 1 and 5 per unit for country 2, with corresponding environmental benefits per unit of 5 resp. 2. The polluting emissions are distributed equally over the two countries as in the case of the Greenhouse effect. In this case, the matrix and the corresponding pay-off diagram are given in Table 5.4 and Figure 5.6 respectively (the game will be referred to as example game I)[36]:

		Player 2	
		D	C
Player 1	D	$(0^*, 0^*)$	$(5, -3)$
	C	$(-1, 2)$	$(4, -1)$

Table 5.4: *Pay-off matrix of Example Game I (* stands for Nash equ.)*

Note that the non-cooperative equilibrium is $(0, 0)$, so that all elements in the shaded area of the pay-off graph can be sustained as an equilibrium outcome of the infinite repetition of the game with low enough values for ρ. Note also the crucial point that the sum of pay-offs is maximised in $(4, -1)$. However, this point is not inside the shaded area and can hence not be sustained as long run equilibrium.

---

[35]Both McGinnis (1986) and Folmer, van Mouche & Ragland (1991; example) make the same point but stress linkage where the cooperative outcome is $(f_1; u_2)$ in one game and (u_1, f_2) in the other.

[36]In the pay-off diagram, $m_1 = 4$, because joint cooperation gives two units of benefit (i.e. 10) and one unit of costs (i.e. 6) for country 1, etc.

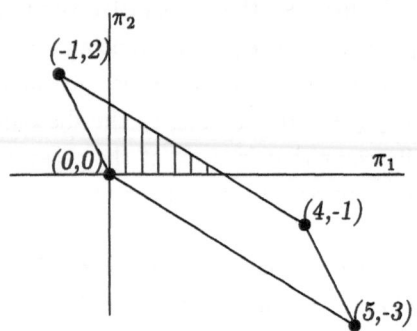

Figure 5.6: *Pay-off space of Example Game I*

Therefore, in order to sustain joint cooperation, the game will be **linked** with a 'roughly offsetting' game. This will be elaborated shortly. First, however, some more concepts and assumptions will be introduced.

Assume that $n_1 + n_2 < u_1 + f_2$, $u_2 + f_1 < m_1 + m_2$. This implies that, the more abatement activities are undertaken, the better it is for joint pay-offs. Hence, cooperation maximises joint pay-offs. This makes (m_1, m_2) the natural **focal point**, at least for an objective outsider.

Given the all-or-nothing framework (D or C), two types of players can be distinguished: those who prefer unilateral abatement to no abatement at all, and those who are not willing to go ahead with unilateral actions[37]. Following Martin (1992-a), the first type is referred to as **strong** and the second as **weak**[38].

Given the focus on asymmetric games where the full cooperative equilibrium cannot be sustained as it is Pareto dominated by the non-cooperative equilibrium for one player. There are two possible types of games with pure strategies that fit in this framework: asymmetric prisoner's dilemma-games and suasion games[39]. Both games will be elaborated here.

---

[37]I do not consider situations in which players are indifferent.

[38]Note that 'strong' players are not necessarily 'dominant' in the game-theoretic sense: 'strong' only says that the player prefers the strategy C when the other player chooses D.

[39]If mixed strategies is allowed for, asymmetric chicken games and other games would fit in the framework. In fact, as stressed Carraro & Siniscalco (1991) and Parson & Zeckhauser (1992) among others, they might be very relevant in environmental games.

Asymmetric Prisoner's Dilemma Games   The example given above in Table 5.4 is an asymmetric prisoner's dilemma game. Its basic characteristic is that non-cooperation is the dominant strategy (and hence, both players are weak). Given the asymmetry, it follows without loss of generality that $n_1 < m_1$ and $n_2 > m_2$.

It seems that in international environmental problems, this game gives an apt description of many situations. For instance, in the case of the Rhine Treaty, for a long time a stalemate position persisted because France was not willing to reduce its discharges. In fact, it was not until the Netherlands and others agreed to offer compensatory payments to France, that the Treaty could be signed (Folmer, van Mouche & Ragland, 1991, p.2)

Suasion Games   In suasion games (see Martin, 1992-b), at least one of the two parties is 'strong' and the other party will in this situation prefer to be free-rider. Hence, the possible equilibrium pay-offs are (u_1, f_2) or (f_1, u_2). Given the assumptions[40], it follows in the former case that , $f_2 > m_2$ and $u_1 < m_1$.

To illustrate this type of game, an example similar to Figure 5.6 is worked out, with constant costs and benefit per unit of abatement. For country 1, these are 4 resp. 5 and for country 2, these are 5 and 2. The pay-off matrix and space are depicted in Table 5.5 and Figure 5.7 respectively (the game will be referred to as example game 2):

Player 2

		D	C
Player 1	D	$(0,0)$	$(5,-3)$
	C	$(1^*,2^*)$	$(6,-1)$

Table 5.5: *Payoff matrix of Example Game II (* stands for Nash equ.)*

In this game, the one shot non-cooperative equilibrium is $(1,2)$, where player 2 is a free rider. Note, that the only individually rational outcome of an infinite repetition of this game is $(1,2)$ in every round of the game, as no other equilibrium will give player 2 at least pay-off 2, that he/she can achieve by free riding. Hence, there are no credible threats within the infinitely repeated game that can sustain the cooperative equilibrium. Note also that joint welfare is maximised in $(6,-1)$.

---

[40]First of all, $f_2 > m_2$, given the definition of a Nash equilibrium. Under the assumption of $n_1 + n_2 < u_1 + f_2$, $u_2 + f_1 < m_1 + m_2$, it follows than automatically that $u_1 < m_1$.

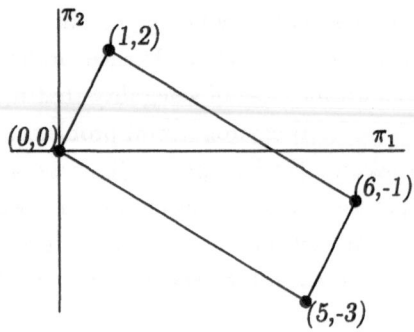

Figure 5.7: *Pay-off matrix of Example Game II*

Again, see how linkage with a second game that is a 'roughly offsetting' mirror image of the first one can overcome the Pareto inferior free riding.

An example of a suasion game is again the ozone case. Cooperation between the North and the South failed in the Montreal Protocol and the North went ahead unilaterally. Major developing countries refused to sign, as it seemed that they would be worse off by cooperating than by free-riding[41]. In fact, it was not until the London Amendments in 1990 where CFC-reduction policies were linked to technology transfers to the South that joint cooperation became feasible.

5.3.3   Linkage in Two-By-Two Games

Assume that there is an exact mirror image of the game in Table 5.4[42] and that both issues will be discussed for joint settlement. Game-theoretically, this means that the controllable subspace is enlarged and the loss function is altered[43]. The resulting four-by-four pay-off matrix and corresponding pay-off space are given in Table 5.6 and Figure 5.8 respectively.

In this game, the objective focal point of $(3, 3)$ is now Pareto dominant vis-à-vis

---

[41]In my view, they had good reason to be free riders, as they could well claim that the ozone problem was caused by 50 years of CFC use in the North.

[42]That means: $n_1 = n_2 = 0$, $m_1 = -1$, $m_2 = 4$, $f_1 = 2$, $f_2 = 5$, $u_1 = -3$ and $u_2 = -1$.

[43]As mentioned in the introduction, one might wonder why these issues were considered separately in the first place. It was argued that there may be political and/or administrative reasons.

Player 2

	$N, \bar{N}$	$N, \bar{C}$	$C, \bar{N}$	$C, \bar{C}$
$(N, \bar{N})$	$(0^*, 0^*)$	$(2, -1)$	$(5, -3)$	$(7, -4)$
$(N, \bar{C})$	$(-3, 5)$	$(-1, 4)$	$(0, 0)$	$(4, 1)$
$(C, \bar{N})$	$(-1, 2)$	$(0, 0)$	$(4, -1)$	$(6, -2)$
$(C, \bar{C})$	$(-4, 7)$	$(-2, 6)$	$(1, 4)$	$(3, 3)$

Player 1 (labels $(N, \bar{N})$, $(N, \bar{C})$, $(C, \bar{N})$, $(C, \bar{C})$ on rows)

Table 5.6: *Pay-off matrix of Example Game III (* stands for Nash equ.)*

the non-cooperative equilibrium $(0, 0)$ and can hence be sustained as equilibrium. Generally, we can say that linkage between an asymmetric prisoner's dilemma game or a suasion game with any other asymmetric prisoner's dilemma game or suasion game makes the full cooperative solution an equilibrium of infinite repetition of the combined game as long as the pay-off of full cooperation in the combined game is larger for each of the players than the sum of the pay-offs of playing the one-shot Nash equilibrium of each of the games.

This can be seen trivially by noting that both an asymmetric prisoner's dilemma game and a suasion game have a unique Nash equilibrium. Hence, a simultaneous play of any pair of such games in the assumed absence of synergy effects has the Nash equilibria of the separate games as equilibrium. With this equilibrium as threat point, any strategy of the infinite repetition of the linked games can be supported as an equilibrium strategy using, for instance, an infinite period trigger strategy because the corresponding pay-offs for each of the players are individually rational, as was assumed (Folk Theorem).

The formulation above of the Folk Theorem uses infinite period punishment as trigger strategies to deter deviation from the equilibrium path. The problem is that these punishments seem to be very severe and may be unstable. This is so, because a cheater might convince the other player to forgive him/her for having reneged, as that is in the interest of that other player to restore cooperation. As elaborated in Section 4.5, the concept of renegotiation proofness has been introduced precisely to avoid such instability[44]. Note that the 'penance strategy' outlined in Section 4.5 is just the familiar **tit-for-tat strategy** (Axelrod, 1984). In what follows, this strategy, as a special case of renegotiation-proof equilibria, is focused on.

---

[44]In this section an example was given and possible continuation plays were formulated.

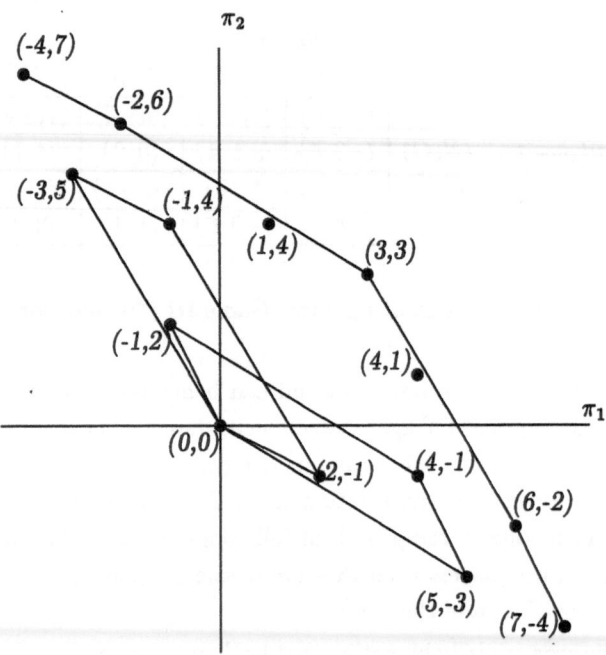

Figure 5.8: *Pay-off matrix of Example Game III*

Note that restrictions need to be placed on linkage between games to guarantee that tit-for-tat strategies can sustain cooperation. In the example at the beginning of this section, an exact mirror image of the first game was taken as second game. In real life, such mirror image games will only by found by a fluke. At the same time, the idea of (specific) reciprocity would urge us to find 'exchanges of roughly equivalent value' (Keohane, 1986, p.4). This idea can be exemplified as follows: "the man who gives a dinner party does not bargain with his guests about what they will do for him in return, but "he expects them not simply to ask him for a quick lunch if he has given a formal dinner for them"[45].

---

[45] Example from Peter Blau (taken from Keohane, 1986, p.6; ref. therein).

Formalisation of the concept of **rough equivalence** in asymmetric games is not trivial. Here, I would argue for the following set of definitions:

Definition:
Linkage of two asymmetric games (suasion games or asymmetric prisoner's dilemma games) gives 'exchanges of roughly equivalent value' if every one shot pure strategy of each of the games is **roughly offset** by its **counterpart strategy** in the other game.

Definition:
The **counterpart** strategies of two linked two-by-two games are:
(C, C) in game A with $(\bar{C}, \bar{C})$ in game B;
(N, C) in game A with $(\bar{C}, \bar{N})$ in game B;
(C, N) in game A with $(\bar{N}, \bar{C})$ in game B;
(N, N) in game A with $(\bar{N}, \bar{N})$ in game B;
  Denoting the pay-offs of game A as n_i, m_i, u_i and m_i (see Table 5.3) and of game B as $\bar{n}_i, \bar{m}_i, \bar{u}_i$ and $\bar{m}_i$, it follows that:

Definition: The one shot pure strategies of each of the asymmetric games are **roughly offset** by their counterpart strategies in the other game, if the following pay-off requirements are met:

$$n_1 + \bar{n}_1 \,,\, n_2 + \bar{n}_2 < u_1 + \bar{f}_1 \,,\, \bar{u}_1 + f_1 \,,\, u_2 + \bar{f}_2 \,,\, \bar{u}_2 + f_2 < m_1 + \bar{m}_1 \,,\, m_2 + \bar{m}_2 \quad (5.21)$$

As an alternative to the requirements above, one could argue that the only relevant requirement is that the full cooperative solution is in the interest of both players (individually rationality). Here, however, an attempt is made to formalise the idea of 'roughly-offsetting' games, as a way of contributing to clarify the concepts used by political scientists is this field.

  The intuition behind these (weak) requirements are to make the counterpart strategies even out. Given the focus here on asymmetric games and given the assumption of that 'the more abatement activities are undertaken, the better it is for joint pay-offs' (i.e. $n_1 + n_2 < u_1 + f_2$, $u_2 + f_1 < m_1 + m_2$), the idea of 'evening out' is that the sum of the pay-offs of the counterpart strategies is lowest for joint non-cooperation and highest for joint cooperation. Graphically, this result can be interpreted easily by normalising (without loss of generality): $n_1 = \bar{n}_1 = n_2 = \bar{n}_2 = 0$.

  The definition above then says that the vector sum of the pay-offs of the counterpart strategies are all in the rectangular with coordinates $\{(0, 0), (0, m_2 + \bar{m}_2), (m_1 +$

$\bar{m}_1, \bar{m}_2 + \bar{m}_2), (m_1 + \bar{m}_1, 0)\}$. This means that the cooperative solution is the one with highest joint pay-off.

This gives the following theorem:

Theorem 5.3   (Proof is given Appendix C.2):
Linkage of two roughly offsetting suasion games satisfying the assumptions given above makes infinite repetition of the full cooperative solution the Pareto optimal and joint pay-off maximisation equilibrium. This equilibrium can be sustained by the following tit-for-tat continuation strategy II, which is renegotiation proof for low enough values of the discount rate[46]:

$$\Pi = \begin{cases} \pi_1 & \text{always play } (C, C; \bar{C}, \bar{C}) \\ \pi_2 & \text{play } (C, D; \bar{C}, \bar{C}) \text{ at t=0, otherwise play } (C, C; \bar{C}, \bar{C}); \\ \pi_3 & \text{play } (C, C; \bar{D}, \bar{C}) \text{ at t=0, otherwise play } (C, C; \bar{C}, \bar{C}); \end{cases} \qquad \square$$

The idea is to consider a whole range of possible linkage opportunities, to check whether they are roughly offsetting and then to consider which of the remaining candidates is the best-suited one. This depends on what would seem 'fair' to both parties.

The linkage of suasion games seems to be particularly interesting, because punishment after defection takes only place in one of the two games. Take as an example the two roughly offsetting suasion games A and B given in Table 5.7 and Table 5.8.

Linking games A and B, the pay-off matrix and space as given in Table 5.9 and Figure 5.9 resp. apply. It can be checked that the requirements are indeed satisfied. Notice that a deviation by player II in game A will be answered with $(C, \bar{D})$ by player I. Thus, player I keeps playing his cooperative strategy in game A, but will punish by free riding in game B. This is crucial. The penance strategy that sustains $(3, 4)$ as cooperative equilibrium could be referred to as **tit-for-roughly-tat**, because reneging in one game leads to punishment in the roughly offsetting other game. For example, if player II played $(D, \bar{C})$, gaining 2 units by free riding in game A, player I would go to $(C, \bar{D})$ and hence, he would punish by free riding in game B. Player II can only restore cooperation by going to $(C, \bar{C})$. The continuation pay-offs after $(C, D; \bar{C}, \bar{C})$ are $(4, 1)$ followed by repetition of $(3, 4)$. This is not Pareto dominated by renegotiation an immediate return to $(3, 4)$ without punishment for low enough values of the discount rate.

---

[46]Assumed is here, without loss of generality, that the one-shot Nash equilibrium is (C, N) and in game II $(\bar{N}, \bar{C})$, denoted as $(C, N; \bar{N}, \bar{C})$

Player 2

	D	C
Player 1 D	$(0,0)$	$(3,-1)$
C	$(1^\star, 2^\star)$	$(3,0)$

Table 5.7: *Game A*

Player 2

	D	C
Player 1 D	$(0,0)$	$(1^\star, 1^\star)$
C	$(-2,3)$	$(0,4)$

Table 5.8: *Game B*

Player 2

	$N,\bar{N}$	$N,\bar{C}$	$C,\bar{N}$	$C,\bar{C}$
$(N,\bar{N})$	$(0,0)$	$(1,1)$	$(3,-1)$	$(4,0)$
$(N,\bar{C})$	$(-2,3)$	$(0,4)$	$(1,2)$	$(3,3)$
$(C,\bar{N})$	$(1,2)$	$(2,3)$	$(3,0)$	$(4,1)$
$(C,\bar{C})$	$(-1,5)$	$(1,6)$	$(1,3)$	$(3,4)$

Table 5.9: *Pay-off matrix of the Linked Games A & B*

Translating this in terms of international negotiations on the Greenhouse effect, this means that the ROECD could go ahead with emission reductions, even if the US seems to free ride on this issue. But in reaction, the ROECD can punish with non-cooperative behaviour on international security issues, e.g. within NATO. If the two issues are indeed 'roughly offsetting', the US might by induced to rethink its Climate Change Policy.

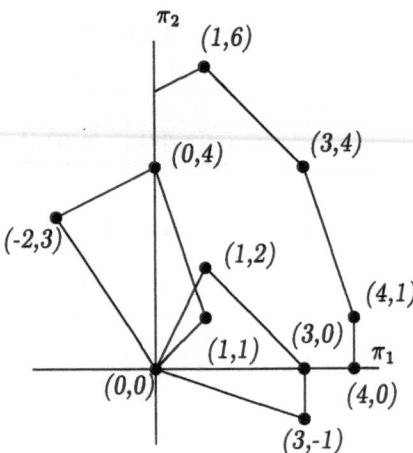

Figure 5.9: *Pay-off matrix of Linked Game A & B (* stands for Nash equ.)*

In the case of the asymmetric Prisoner's dilemma, a similar theorem can be stated:

Theorem 5.4   (Proof is given Appendix C.2):

Linkage of two roughly offsetting asymmetric prisoner's dilemma games satisfying the assumptions given above makes infinite repetition of the full cooperative solution the Pareto optimal and maximum joint pay-off equilibrium. This equilibrium can be sustained by the following tit-for-tat strategy profile Π, which is renegotiation proof[47]:

$$\Pi = \begin{cases} \pi_1 & \text{always play } (C,C;\bar{C},\bar{C}) \\ \pi_2 & \text{play } (C,D;\bar{C},\bar{D}) \text{ at } t=0, \text{ otherwise play } (C,C;\bar{C},\bar{C}); \\ \pi_3 & \text{play } (D,C;\bar{D},\bar{C}) \text{ at } t=0, \text{ otherwise play } (C,C;\bar{C},\bar{C}); \end{cases} \qquad \square$$

Note that here, upon violation of the cooperative agreement in one (or both) of the two games, punishment by the 'victim' will take place by using non-cooperative strategies in both games. Translating this back to real life situations in which countries are involved in negotiations on several issues simultaneously, linkage of these issues means using the threat of non-cooperative behaviour in all of these

---

[47]$(C,D;\bar{C},\bar{D})$ means playing (C,D) in game A and $(\bar{C},\bar{D})$ in game B.

issues as leverage in order to sustain cooperation on one particular issue. This idea
will be further worked out in the more general framework of differential games.

5.3.4 Linkage in Differential Games

As elaborated in Section 4.4, cooperation in differential games cannot be sustained
using trigger strategies and renegotiation proof strategies, if cooperation is not
'agreeable' in the sense described in Section 4.4. Linkage might resolve the problem.
It may also be that the only 'agreeable' forms of cooperation are suboptimal from a
welfare point of view, and that the socially optimal solution may become agreeable
once linkage of the asymmetric game with a mirror image game is considered.

An interesting example of such a form of linkage is the North Pacific Seal Treaty,
aimed at forbidding a specific form of harvesting seals, as described by Barrett
(1991, OECD, Ch.3). "The 1957 Interim Convention, like the 1911 Treaty that
preceded it, strictly limits the wasteful practice of pelagic sealing — the taking of
seals at sea — but does not limit harvests by land. Since the United States and
USSR owned the important fur seal breeding grounds, the other two sealing nations,
Canada and Japan, would only lose by signing the agreement to prohibit pelagic
sealing. The Treaty therefore requires that a portion of US and USSR harvest
be transferred to Canada and Japan. The 1957 Treaty requires that the US and
USSR give Canada and Japan 15 per cent of their annual harvest of pelts."[48] This
means that the economically efficient situation is the one where only the US and
the USSR would be harvesting on their own territory. Canada and Japan would
discontinue their seal-activities. This, however, makes the latter two countries
worse off. Probably, other forms of cooperation, economically less efficient though,
could have been possible where every actor was better off than in the situation
without an agreement. But by linkage of the decision of where to harvest with
in-kind transfers, a socially superior solution could be obtained[49]. This seal-game
is typically a differential game with the stock of seals being the state variable and
the harvest per year the control variable for each player.

Interestingly, a trigger strategy is built into the Treaty in order to prevent
violation, as noted by Barrett (OECD, Ch. 3, 1991). "Article 12 of the treaty

---

[48]Barrett, OECD, Ch.3, 1991, p.128.

[49]One may argue that the in-kind transfers are just a straightforward alternative of monetary
side-payments and that this is not a real situation of linkage. On the other hand, one might see
the decision of where to harvest in the absence of side-payments as one game and the allocation
of the harvest among the Treaty members as another game.

states that in the event of a violation, and provided signatories cannot agree to 'remedial measures', any signatory may give notice of its intention to terminate the agreement, and that the whole of the agreement will then be terminated nine months later. Since signatories must be better off with the agreement that without it, this article provides strong incentives for full compliance."[50]

In the remainder of this section, the use of trigger strategies and WRP strategies will be formally worked out for the three cases described above. Assume for this purpose that there are two games, 1 and 2. For simplicity, the games have no common variables and there are no synergy effects in combining their pay-offs. The vector of state variables of the two games are $x(1,.)$ and $x(2,.)$. Define $x(.)$ as $[x(1,.)\ x(2,.)]$. The Pareto optimal control for game 1 is (compare Section 4.4: $u^{\star}_{\alpha_1}(1,.)$. The corresponding continuation pay-off at time t is $V^{c}_{i}(1,\alpha_1,t,x(1,t),u^{\star}_{\alpha_1}(1,.))$. Similarly, the optimal feedback Nash control is $u^{\star}(1,.)$ The feedback Nash continuation pay-off for game 1 at time t is $V^{f}_{i}(1,t,x(1,t),u^{\star}(1,.))$. An exact analogue exists for game 2. With these notations, the joint pay-offs of linking games 1 and 2 can be defined as:

$$V^{c}_{i}(\alpha,t,x(t),u^{\star}_{\alpha}(.)) = V^{c}_{i}(1,\alpha_1,t,x(1,t),u^{\star}_{\alpha_1}(1,.)) + V^{c}_{i}(2,\alpha_2,t,x(2,t),u^{\star}_{\alpha_2}(2,.))$$
$$V^{f}_{i}(t,x(t),u^{\star}(.)) = V^{f}_{i}(1,t,x(1,t),u^{\star}(1,.)) + V^{f}_{i}(2,t,x(2,t),u^{\star}(2,.))$$

Concentration will crucially be on situations where linkage is necessary for cooperation given a certain weight α between the welfare functions of the two players. That is, assume that game 1 is not agreeable for player a and game 2 is not agreeable for player b. Linkage could be successful if an 'agreeable' control vector $u^{\star}(.)$ $= [u^{\star}(1,.)\ u^{\star}(2,.)]$ exists (for a definition of 'agreeable controls', see Section 4.4).

In the last section, 'roughly offsetting' strategies were introduced in infinitely repeated two-by-two games. A straight generalisation of this concept for stage games with continuous strategy space, such as differential games, is not trivial. It would mean that for every possible combination of solutions of the individual games an under and upper constraint for the summed pay-off would have to be defined. Therefore, only the necessary requirements for individual rationality and time-consistency are stated for players a and b and games 1 and 2, so that these games can be sustained when treated together, where they could not when treated separately. Note that, as before, the focus is on outcomes at the Pareto Frontier, as other pay-offs may not have uniquely corresponding controls. Besides, only linear-quadratic games are considered, as in Section 4.4.

---

[50]Barrett, OECD, Ch.3, 1991, p. 132

The necessary requirements are for the individual pay-offs of each game:

$$V_a^c(1, \alpha_1, t, x(1,t), u_{\alpha_1}^*(1,.)) \quad > \quad V_a^f(1, t, x(1,t), u^*(1,.))$$
$$V_a^c(2, \alpha_2, t, x(2,t), u_{\alpha_2}^*(2,.)) \quad < \quad V_a^f(2, t, x(2,t), u^*(2,.))$$
$$V_b^c(1, \alpha_1, t, x(1,t), u_{\alpha_1}^*(1,.)) \quad < \quad V_b^f(1, t, x(1,t), u^*(1,.))$$
$$V_b^c(2, \alpha_2, t, x(2,t), u_{\alpha_2}^*(2,.)) \quad > \quad V_b^f(2, t, x(2,t), u^*(2,.))$$

And for the joint pay-offs of the linked game:

$$V_a^c(\alpha, t, x(t), u_\alpha^*(.)) \quad > \quad V_a^f(t, x(t), u^*(.))$$
$$V_b^c(\alpha, t, x(t), u_\alpha^*(.)) \quad > \quad V_b^f(t, x(t), u^*(.))$$

These expressions have to hold for all possible values of $x(t)$ along the Pareto efficient trajectory of $x(.)$.

With this necessary requirements stated, trigger strategies $\langle \hat{u}_\alpha(.) \rangle$ and infinitesimal δ-continuation strategies (for the WRP case) $\langle \hat{\pi}_\alpha(.) \rangle$ can be defined as in Section 4.4. Finally, natural extension of the Theorem and Conjecture of Section 4.4 gives the conditions under which trigger and WRP strategies form a Pareto optimal subgame perfect equilibrium of the linked games 1 and 2.

Crucial in the statements above is that combining the two games does not have synergy effects. Therefore the pay-offs can just be added. In this way, the linkage has created a leverage that induces the players to respect the joint cooperative agreement, where they would have an incentive to renege upon the agreement in the individual games.

The analysis here is rather straightforward, as the weights α are taken to be fixed. If these were allowed to change, it would be much more complicated. Another problem is that for different α the final linked cooperative pay-offs might be more 'fairly' distributed between the actors than for other values of α. This has not been dealt with here. Typically, from an efficiency point of view, the closer the α to the socially optimal one, the better. From an equity point of view, linkage is most useful of two games of which the pay-offs corresponding to the preferred values of α are as close mirror images as possible.

It seems obvious that schemes as described above could not directly be applied in agreements between countries. At the same time, the North Pacific Seal Treaty shows that there are ways of approaching punishment strategies that do work in practice and that sustain cooperation to the benefit of all involved.

5.4   Summary and Conclusions

This chapter has dealt with cooperation in a second-best world. Especially, two such instances have been investigated: technology transfers and issue linkage. First, technology transfers in emission reduction were discussed using a two country dynamic game model. In this framework, countries could basically decide whether to invest in emission reduction at home or abroad, with efficiency costs outside the home country. At the same time, abatement could be lower abroad which would compensate for these costs. Summarising, the use of technology transfers in emission reduction could be a promising option both in non-cooperative and in cooperative settings. Given certain conditions, this is true even if the receiving country cannot commit itself to its previously announced abatement policy once technology transfers are introduced.

Next, we looked at issue linkage. This refers to the combined negotiation of two different items at the same time with the aim of joint settlement. In game-theoretic terms this means that the controllable subspace is enlarged. Special focus was the linkage of roughly-offsetting asymmetric games, where the mere repetition of each game individually over time did not establish a social optimum. Linkage created or expanded a zone of possible agreement between the parties. In the repeated two-by-two framework, we analysed two games in detail: the asymmetric prisoner's dilemma and the asymmetric suasion game. It was shown that the social optimum could be sustained using trigger strategies or renegotiation-proof strategies. The same was done subsequently in a differential game framework. The most interesting result, in my opinion, is the linkage of asymmetric suasion games, where players typically have each one issue that they care about and one that they don't. In this case, tit-for-tat strategies can be constructed where punishment after defection takes only place in the game that the punisher does not care about, and the reneger feels very strongly about. Such strategies are shown to be renegotiation proof.

In this last chapter, I have discussed two issues (technology transfers and linkage) that are typically not covered in the game-theoretic literature on international environmental policy, but that seem to be very important in actual negotiations. It appears to me that environmental economists using game theory do not address the crucial issue of how to get from 'here' (Nash non-cooperative) to 'there' (Pareto optimum). The analysis of these two issues are attempts to deal with the question of how to use with second best deals to get negotiations going when the grand optimum optimorum is not (yet) available.

Appendix A

A.1 Local Stability Analysis for Section 3.3.1

Write the Jacobian matrix of the linearised system evaluated at the steady state $(P^\infty, T^\infty, \psi^\infty, \xi^\infty)$ as:

$$J = \begin{bmatrix} \rho I - A & B \\ C & D \end{bmatrix} \qquad (A.1)$$

with:

$$A = \begin{bmatrix} \mathcal{H}_{P\psi} & \mathcal{H}_{P\xi} \\ \mathcal{H}_{T\psi} & \mathcal{H}_{T\xi} \end{bmatrix} \qquad (A.2)$$

The elements of the Jacobian sub-matrix A are[1]:

$$
\begin{aligned}
\mathcal{H}_{P\psi} &= -\alpha + \xi Y_{PF} F_\psi & < 0 \\
\mathcal{H}_{P\xi} &= \xi Y_{PF} F_\xi & < 0 \\
\mathcal{H}_{T\psi} &= \xi Y_{TF} F_\psi & > 0 \\
\mathcal{H}_{T\xi} &= -d_T + \xi Y_{TF} F_\xi + Y_T & > 0
\end{aligned}
$$

Likewise, B is:

$$B = \begin{bmatrix} -\mathcal{H}_{PP} & -\mathcal{H}_{PT} \\ -\mathcal{H}_{TP} & -\mathcal{H}_{TT} \end{bmatrix} \qquad (A.3)$$

---

[1] $\mathcal{H}_{T\xi} > 0$ if d_T not too large.

where the elements of the Jacobian sub-matrix B are[2]:

$$
\begin{aligned}
-\mathcal{H}_{PP} &= -\xi Y_{PF}F_P - \xi Y_{PP} &\geq 0 \\
-\mathcal{H}_{PT} &= -\xi Y_{PF}F_T - \xi Y_{PT} &> 0 \\
-\mathcal{H}_{TP} &= -\xi Y_{TF}F_P - \xi Y_{TP} &> 0 \\
-\mathcal{H}_{TT} &= -\xi Y_{TF}F_T - \xi Y_{TT} &\geq 0
\end{aligned}
$$

And C is

$$
C = \begin{bmatrix} \mathcal{H}_{\psi\psi} & \mathcal{H}_{\psi\xi} \\ \mathcal{H}_{\xi\psi} & \mathcal{H}_{\xi\xi} \end{bmatrix}
\tag{A.4}
$$

with elements of the Jacobian sub-matrix C:

$$
\begin{aligned}
\mathcal{H}_{\psi\psi} &= \gamma F_\psi &> 0 \\
\mathcal{H}_{\psi\xi} &= \gamma F_\xi &> 0 \\
\mathcal{H}_{\xi\psi} &= Y_F F_\psi &> 0 \\
\mathcal{H}_{\xi\xi} &= Y_F F_\xi - C_\xi &> 0
\end{aligned}
$$

Finally, D is:

$$
D = \begin{bmatrix} \mathcal{H}_{\psi P} & \mathcal{H}_{\psi T} \\ \mathcal{H}_{\xi P} & \mathcal{H}_{\xi T} \end{bmatrix}
\tag{A.5}
$$

where the elements of the Jacobian sub-matrix D^3 are:

$$
\begin{aligned}
\mathcal{H}_{\psi P} &= \gamma F_P - \alpha &< 0 \\
\mathcal{H}_{\psi T} &= \gamma F_T &> 0 \\
\mathcal{H}_{\xi P} &= Y_F F_P + Y_P &< 0 \\
\mathcal{H}_{\xi T} &= -d_T + Y_F F_T + Y_T &> 0
\end{aligned}
$$

Note that the matrices B and C are symmetric and positive definite. This can be shown by observing that: $F_P = -\frac{Y_{FP}}{Y_{FF}}$, $F_T = -\frac{Y_{FT}}{Y_{FF}}$, $F_\psi = -\frac{\gamma}{\xi Y_{FF}}$ and $F_T = -\frac{Y_F}{\xi Y_{FF}}$
Hence, $\mathcal{H}_{TP} = \mathcal{H}_{PT}$; $\mathcal{H}_{\psi\xi} = \mathcal{H}_{\xi\psi}$.
Also, note that $-\mathcal{H}_{PP} \geq 0$ and that $\mathcal{H}_{PP}\mathcal{H}_{TT} - \mathcal{H}_{PT}\mathcal{H}_{TP} \geq 0$ due to our concavity assumptions.
Likewise, $\mathcal{H}_{\psi\psi} > 0$ and $\mathcal{H}_{\psi\psi}\mathcal{H}_{\xi\xi} - \mathcal{H}_{\psi\xi}\mathcal{H}_{\xi\psi} > 0$. This means that B and C are symmetric and positive (semi) definite.

---

[2]Due to concavity assumption that $Y_{PP}Y_{FF} - Y_{PF}^2 \geq 0$, and that $Y_{FF}Y_{TT} - Y_{FT}^2 \geq 0$, it follows that resp. $-\mathcal{H}_{PP} \geq 0$ and $-\mathcal{H}_{TT} \geq 0$. Note that in the case that $Y(e(F,T),P) = e(F,T).f(P)$ where $e(.)$ is a constant returns to scale Cobb-Douglas function, this last second order product is zero.

[3]$\mathcal{H}_{\xi T} > 0$ if d_T not too large.

In the same way, $\mathcal{H}_{P\psi} = \mathcal{H}_{\psi P}$; $\mathcal{H}_{P\xi} = \mathcal{H}_{\xi P}$; $\mathcal{H}_{\psi T} = \mathcal{H}_{T\psi}$ and $\mathcal{H}_{\xi T} = \mathcal{H}_{T\xi}$ so that $-A^T = D$.

As the matrices B and C are symmetric and positive definite, the equilibrium solution of the dynamical system 3.40 is a saddle point if the rate of time preference ρ is sufficiently low[4] (Brock & Scheinkman (1976)) and the steady state is locally asymptotically stable[5].

A.2   Local Stability Analysis for Section 3.4.1

As in Section 3.3.1, the Jacobian of System 3.65 can be written as:

$$J = \begin{bmatrix} \rho I - A & B \\ & C & D \end{bmatrix}$$

with:

A:
$$\begin{aligned}
\mathcal{H}_{K\xi} &= -d_K + \xi Y_{KF} F_\xi + Y_K \\
\mathcal{H}_{H\xi} &= -Y_H - \xi Y_{HF} F_\xi \\
\mathcal{H}_{Kx} &= 0 \\
\mathcal{H}_{Hx} &= -d_H
\end{aligned}$$

B:
$$\begin{aligned}
-\mathcal{H}_{KK} &= -\xi Y_{KF} F_K - \xi Y_{KK} \\
-\mathcal{H}_{HK} &= -\xi Y_{HF} F_K - \xi Y_{HK} \\
-\mathcal{H}_{KH} &= -\xi Y_{KF} F_H - \xi Y_{KH} \\
-\mathcal{H}_{HH} &= -\xi Y_{HF} F_H - \xi Y_{HH}
\end{aligned}$$

C:
$$\begin{aligned}
\mathcal{H}_{\xi\xi} &= Y_F F_\xi - C_\xi - I_{2x} \\
\mathcal{H}_{\xi x} &= -I_{2x} \\
\mathcal{H}_{x\xi} &= \theta I_2^{\theta-1} I_{2\xi} \\
\mathcal{H}_{xx} &= \theta I_2^{\theta-1} I_{2x}
\end{aligned}$$

---

[4]To see this, define

$$Q = \begin{bmatrix} O & I \\ -I & 0 \end{bmatrix} \tag{A.6}$$

Then, one can show that $QJQ^{-1} = -J^T + \rho I$. Now assume that η is an eigenvalue of J with x being the corresponding eigenvector. Then $QJx = Q\eta x = \eta Qx = -J^T Qx + \rho IQx$. This means that $-\eta + \rho$ is an eigenvalue as well. Hence, for ρ not too high, η and $(\rho - \eta)$ have opposite signs (c.f. Van der Ploeg & Withagen (1991) and Musu (1992)).

[5]See Feichtinger & Hartl (1986) for an alternative approach.

$$D: \qquad \begin{aligned} \mathcal{H}_{\xi K} &= -d_H + Y_F F_K + Y_K \\ \mathcal{H}_{\chi K} &= 0 \\ \mathcal{H}_{\xi H} &= Y_F F_H + Y_H \\ \mathcal{H}_{\chi H} &= -d_H \end{aligned}$$

Due to the concavity assumptions made on $Y\{e(F,K,H),P\}$, the second deriva-
tive of $\mathcal{H}$ with respect to K of the sub-matrix B is known to be negative ($-\mathcal{H}_{KK}$
¿ 0). Likewise, $\mathcal{H}_{KK}\mathcal{H}_{HH} - \mathcal{H}_{HK}\mathcal{H}_{KH} < 0$.

Note also that B is symmetric as $\frac{F_K}{F_H} = \frac{Y_{FK}}{Y_{FH}}$. Therefore, B is a positive definite
symmetric matrix.

Similarly, $\mathcal{H}_{\xi\xi} > 0$ for submatrix C and $\mathcal{H}_{\xi\xi}\mathcal{H}_{\chi\chi} - \mathcal{H}_{\chi\xi}\mathcal{H}_{\xi\chi} < 0$. The latter is
straightforward as $Y_F F_\xi > 0$ and $C_\xi < 0$. Therefore, C is positive definite. As
$\frac{I_{2\chi}}{I_{2\xi}} = -\theta I_2^{\theta-1}$, C is also symmetric.

By the same logic, $A^T = D$ (as $\frac{F_\xi}{F_K} = \frac{Y_F}{\xi Y_{FK}}$ and $\frac{F_\chi}{F_H} = \frac{Y_F}{\xi Y_{FH}}$).

Hence the equilibrium solution of the dynamical system 3.65 is a saddle point if
the rate of time preference ρ is sufficiently low[6] and the steady state of the restricted
dynamical system is locally asymptotically stable.

---

[6]see corresponding analysis in Section 3.3.1

Appendix B

B.1 Routine to Calculate Riccati Coefficients

Miller & Salmon (1985) suggest a procedure for calculating the strategies in the linear-quadratic feedback case. It runs as follows[1]:

1. Express the controls as linear functions of the state variables with initial guesses of the coefficients;

2. Derive the first order conditions of the feedback case with unknown values of the Riccati coefficients;

3. Substitute the control variables out of the first order conditions; this gives a system of differential equations in the states and co-states alone;

4. Solve this system, conditional on the unknown values of the Riccati coefficients; one can now express the co-states as functions of the states with the aid of the eigenvectors of the system;

5. Rewrite the static first order conditions $\frac{\partial \mathcal{H}_i}{\partial u_i} = 0$ as functions of x and u alone using the expressions given in [3];

6. Use the functions in [5] to update [1];

7. Repeat [2]-[6] till convergence has been reached.

---

[1]This procedure can be carried out using Mark-2. This is a version of SADDLEPOINT (Austin & Buiter, 1982) adapted by Mark Salmon to allow for an iterative calculation. The updating is to be done manually. For a short description of how to use this programme, see Cesar & Della Posta (1991)

This alternative procedure leads to the Riccati coefficients. The advantage of the routine is that it may be quite tedious to figure out which of the multiple solutions of the Riccati coefficients, calculated with the coupled system of Riccati equations is the one for which the transversality conditions are met.

B.2   Example Worked Out

To illustrate the techniques given in Section 4.2, an example[2] is worked out for the linear-quadratic case with the following values of the (scalar) parameters in the objective functional (4.2) and the equation of motion (4.3): $A = -1$, $B_i = C_i = D_i = 1$, $E_i = 0$, $c_i = d_i = h_i = 0$.

In the *feedback* case, following Equation 4.7, it is claimed that $V_x = \psi_i = x'S_i^3$. Then, the optimal strategy for player i is: $u_i^* = S_i x$ and following the theorem above, the coupled Riccati differential equations are:

$$0 = -2S_i - \rho S_i - S_i^2 - 2S_i S_j + 1 \tag{B.1}$$

In the *open-loop* case, using Pontryagin's maximum principle, the necessary conditions of the System 4.2– 4.3 are:

$$\dot{x} = -x + u_1 + u_2 \tag{B.2}$$
$$\dot{\psi}_i = \rho \psi_i - x + \psi_i \tag{B.3}$$
$$u_i = -\psi_i \tag{B.4}$$

Using the linear relation $\psi_i = S_i x$, differentiating with respect to time, and substituting in Equations B.2– B.3, gives:

$$-S_i x + S_i u_1 + S_i u_2 = \rho \psi_i - x + \psi_i$$

Substituting u_i and ψ_i out of this equation with the help of the relations $\psi_i = S_i x$ and the first order condition $u_i = -\psi_i$, gives:

$$0 = -2S_i x - \rho S_i x - S_i^2 x - S_i S_j x + x$$

This equation should hold for every value of x, and therefore, the coupled Riccati differential equations for the open-loop case are:

$$0 = -2S_i - \rho S_i - S_i^2 - S_i S_j + 1 \tag{B.5}$$

---

[2]This example is used in de Zeeuw (1991) in a finite time horizon minimisation setting.
[3]Note that with $c_i = d_i = h_i = 0$, it is claimed that $s_i = 0$.

Comparison of Equation B.5 and Equation B.1, shows that the open-loop and the closed-loop systems are indeed different in general.

The derivation for the Riccati equation for the *Pareto* (or *joint optimisation*) case is an exact analogue of the one given above for the open-loop case. The resulting equation is:

$$0 = -2S - \rho S - 2S^2 + 2 \tag{B.6}$$

It is instructive to compare this equation with the Riccati systems of the open-loop and the feedback information pattern for the special case that the two players are identical (i.e. $S_i = S_j \equiv S$).
This gives:

$$
\begin{array}{lllll}
0 & = & -2S - \rho S & - & 3S^2 & + & 1 \qquad & \text{feedback case} \\
0 & = & -2S - \rho S & - & 2S^2 & + & 1 \qquad & \text{open-loop case} \\
0 & = & -2S - \rho S & - & 2S^2 & + & 2 \qquad & \text{Pareto case}
\end{array}
$$

This shows that, in general, the three cases give different results.

B.3 Linear-Quadratic Case Worked Out

The linear-quadratic model described in Section 4.3 above with welfare function:
$U[Y\{e(T,F),P\}-I] = [\omega_1 T_i - \frac{1}{2}\omega_2 T_i^2 + \omega_3 T_i F_i + \omega_4 F_i - \frac{1}{2}\omega_5 F_i^2 - \frac{1}{2}\omega_6 P^2 - \omega_7 I_i - \frac{1}{2}\omega_8 I_i^2]$
and with differential equations $\dot{P} = \gamma \sum_{j=1}^N F_j - \delta P$ and $\dot{T}_i = I_i - d_T T_i$ can be written in matrix notation as follows:

$$\max_{u_i} J_i(.) = \int_0^\infty [\frac{1}{2}u_i' C_i u_i + \frac{1}{2}x' D_i x + x' E_i u_i + c_i' u_i + d_i' x + h_i] e^{-\rho t} dt$$

s.t.

$$\dot{x} = Ax + B_1 u_1 + B_2 u_2$$

with:

$$
A = \begin{bmatrix} -\alpha & 0 & 0 \\ 0 & -d_{T1} & 0 \\ 0 & 0 & -d_{T2} \end{bmatrix} \quad
B_1 = \begin{bmatrix} \gamma & 0 \\ 0 & 1 \\ 0 & 0 \end{bmatrix} \quad
B_2 = \begin{bmatrix} \gamma & 0 \\ 0 & 0 \\ 0 & 1 \end{bmatrix}
$$

$$
D_1 = \begin{bmatrix} -\omega_{61} & 0 & 0 \\ 0 & -\omega_{21} & 0 \\ 0 & 0 & 0 \end{bmatrix} \quad
D_2 = \begin{bmatrix} -\omega_{62} & 0 & 0 \\ 0 & 0 & 0 \\ 0 & 0 & -\omega_{22} \end{bmatrix}
$$

$$
C_i = \begin{bmatrix} -\omega_{5i} & 0 \\ 0 & -\omega_{8i} \end{bmatrix} \quad
c_i = \begin{bmatrix} -\omega_{4i} \\ -\omega_{7i} \end{bmatrix}
$$

$$E_1 = \begin{bmatrix} 0 & 0 \\ w_{31} & 0 \\ 0 & 0 \end{bmatrix} \quad E_2 = \begin{bmatrix} 0 & 0 \\ w_{32} & 0 \end{bmatrix}$$

$$d_1 = \begin{bmatrix} 0 & w_{11} & 0 \end{bmatrix}' \quad d_2 = \begin{bmatrix} 0 & 0 & w_{12} \end{bmatrix}' \quad h_i = 0$$

B.4 Open-Loop Case

For the open-loop case, the first order conditions given in the main text lead to the following steady state conditions for the two-player case:

$$
\begin{align}
D_1 x + E_1 u_1 + d_1 + A' \psi_1 &= \rho \psi_1 \tag{B.7} \\
D_2 x + E_2 u_2 + d_2 + A' \psi_2 &= \rho \psi_2 \tag{B.8} \\
A x + B_1 u_1 + B_2 u_2 &= 0 \tag{B.9} \\
C_1 u_1 + E_1 x + c_1 + B_1' \psi_1 &= 0 \tag{B.10} \\
C_2 u_2 + E_2 x + c_2 + B_2' \psi_2 &= 0 \tag{B.11}
\end{align}
$$

Rewrite Equations B.7 – B.9 as $G_a y = G_b u - g_a$ and Equations B.10 – B.11 as $G_c y = G_d u - g_b$ with y defined as $y \equiv [\, x \; \psi_1 \; \psi_2 \,]'$ and u as $u \equiv [\, u_1 \; u_2 \,]'$. Hence, u can be rewritten as:

$$u = [I - G_d^{-1} G_c G_a^{-1} G_b]^{-1} [G_d^{-1} g_b - G_d^{-1} G_c G_a^{-1} g_a]$$

This gives rise to the following control variables (leaving out the subscript i):

$$F = \frac{\frac{w_3}{d_T}(w_1 - (\rho + d_T) w_7) + ((\rho + d_T) w_8 + \frac{w_2}{d_T}) w_4}{((\rho + d_T) w_8 + \frac{w_2}{d_T})(w_5 + 2\frac{\gamma^2 w_6}{\alpha(\rho + \alpha)}) - \frac{w_3^2}{d_T}} \tag{B.12}$$

$$I = \frac{(w_1 - (\rho + d_T) w_7)(w_5 + 2\frac{\gamma^2 w_6}{\alpha(\rho + \alpha)}) + w_3 w_4}{((\rho + d_T) w_8 + \frac{w_2}{d_T})(w_5 + 2\frac{\gamma^2 w_6}{\alpha(\rho + \alpha)}) - \frac{w_3^2}{d_T}} \tag{B.13}$$

Knowing that at the steady state, $P = \frac{1}{\alpha}\gamma(F_1 + F_2)$ and $T_i = \frac{1}{d_{T_i}} I_i$, the values for P and T_i follow directly from Equations B.12 and B.13.

B.5   Closed-Loop Case

From the theorem on linear-quadratic systems Section 4.2, the following coupled Riccati equations were given:

$$
\begin{aligned}
0_{(n \times n)} &= A'S_i + S_i A - \rho S_i - (S_i B_i + E_i)C_i^{-1}(B_i'S_i + E_i') - \cdots \\
&\quad - (S_j'B_j + E_j)C_j^{-1}B_j'S_i - S_i B_j C_j^{-1}(B_j'S_j + E_j') + D_i \quad (B.14) \\
0_{(n \times 1)} &= A's_i - \rho s_i - (S_i B_i + E_i)C_i^{-1}(B_i's_i + c_i - \cdots \\
&\quad - (S_j'B_j + E_j)C_j^{-1}B_j's_i - S_i B_j C_j^{-1}(B_j's_j + c_j) + d_i \quad (B.15)
\end{aligned}
$$

In the case considered here, $0_{(3 \times 3)}$ resp. $0_{(3 \times 1)}$. Hence, the system has 12 equations for each player. For the specification of the matrices given above, the system of B.14 represents the following 9 equations for player 1[4]:

$$
\begin{aligned}
0 &= -(2\alpha + \rho)\sigma_{11} - \gamma\sigma_{11}\tfrac{-1}{w_{51}}\gamma\sigma_{11} - \sigma_{12}\tfrac{-1}{w_{81}}\sigma_{21} - \gamma\varsigma_{11}\tfrac{-1}{w_{52}}\gamma\sigma_{11} - \\
&\quad -\varsigma_{21}\tfrac{-1}{w_{82}}\sigma_{31} - \gamma\sigma_{11}\tfrac{-1}{w_{52}}\gamma\varsigma_{11} - \sigma_{13}\tfrac{-1}{w_{82}}\varsigma_{31} - w_6 \\
0 &= -\rho\sigma_{12} - \gamma\sigma_{11}\tfrac{-1}{w_{51}}(\gamma\sigma_{12} + w_{31}) - \sigma_{12}\tfrac{-1}{w_{81}}\sigma_{22} - \gamma\varsigma_{11}\tfrac{-1}{w_{52}}\gamma\sigma_{12} - \\
&\quad -\varsigma_{21}\tfrac{-1}{w_{82}}\sigma_{32} - \gamma\sigma_{11}\tfrac{-1}{w_{52}}\gamma\varsigma_{12} - \sigma_{13}\tfrac{-1}{w_{82}}\varsigma_{32} - 0 \\
0 &= -\rho)\sigma_{13} - \gamma\sigma_{11}\tfrac{-1}{w_{51}}\gamma\sigma_{13} - \sigma_{12}\tfrac{-1}{w_{81}}\sigma_{23} - \gamma\varsigma_{11}\tfrac{-1}{w_{52}}\gamma\sigma_{13} - \\
&\quad -\varsigma_{21}\tfrac{-1}{w_{82}}\sigma_{33} - \gamma\sigma_{11}\tfrac{-1}{w_{52}}(\gamma\varsigma_{13} + w_{32}) - \sigma_{13}\tfrac{-1}{w_{82}}\varsigma_{33} - w_6
\end{aligned}
$$

$$
\begin{aligned}
0 &= -\rho\sigma_{21} - (w_{31} + \gamma\sigma_{21})\tfrac{-1}{w_{51}}\gamma\sigma_{11} - \sigma_{22}\tfrac{-1}{w_{81}}\sigma_{21} - \gamma\varsigma_{12}\tfrac{-1}{w_{52}}\gamma\sigma_{11} - \\
&\quad -\varsigma_{22}\tfrac{-1}{w_{82}}\sigma_{31} - \gamma\sigma_{21}\tfrac{-1}{w_{52}}\gamma\varsigma_{11} - \sigma_{23}\tfrac{-1}{w_{82}}\varsigma_{31} - 0 \\
0 &= -(2d_T + \rho)\sigma_{22} - (w_{31} + \gamma\sigma_{21})\tfrac{-1}{w_{51}}(\gamma\sigma_{12} + w_{31}) - \sigma_{22}\tfrac{-1}{w_{81}}\sigma_{22} - \\
&\quad -\gamma\varsigma_{12}\tfrac{-1}{w_{52}}\gamma\sigma_{12} - \varsigma_{22}\tfrac{-1}{w_{82}}\sigma_{32} - \gamma\sigma_{21}\tfrac{-1}{w_{52}}\gamma\varsigma_{12} - \sigma_{23}\tfrac{-1}{w_{82}}\varsigma_{32} - w_2 \\
0 &= -\rho\sigma_{23} - (w_{31} + \gamma\sigma_{21})\tfrac{-1}{w_{51}}\gamma\sigma_{13} - \sigma_{22}\tfrac{-1}{w_{81}}\sigma_{23} - \gamma\varsigma_{12}\tfrac{-1}{w_{52}}\gamma\sigma_{13} - \\
&\quad -\varsigma_{22}\tfrac{-1}{w_{82}}\sigma_{33} - \gamma\sigma_{21}\tfrac{-1}{w_{52}}(\gamma\varsigma_{13} + w_{32}) - \sigma_{23}\tfrac{-1}{w_{82}}\varsigma_{33} - 0
\end{aligned}
$$

$$
\begin{aligned}
0 &= -\rho\sigma_{31} - \gamma\sigma_{31}\tfrac{-1}{w_{51}}\gamma\sigma_{11} - \sigma_{32}\tfrac{-1}{w_{81}}\sigma_{21} - (w_{32} + \gamma\varsigma_{13})\tfrac{-1}{w_{52}}\gamma\sigma_{11} - \\
&\quad -\varsigma_{23}\tfrac{-1}{w_{82}}\sigma_{31} - \gamma\sigma_{31}\tfrac{-1}{w_{52}}\gamma\varsigma_{11} - \sigma_{33}\tfrac{-1}{w_{82}}\varsigma_{31} - w_6 \\
0 &= -\rho\sigma_{32} - \gamma\sigma_{31}\tfrac{-1}{w_{51}}(\gamma\sigma_{12} + w_{31}) - \sigma_{32}\tfrac{-1}{w_{81}}\sigma_{22} - (w_{32} + \gamma\varsigma_{13})\tfrac{-1}{w_{52}}\gamma\sigma_{12} - \\
&\quad -\varsigma_{23}\tfrac{-1}{w_{82}}\sigma_{32} - \gamma\sigma_{31}\tfrac{-1}{w_{52}}\gamma\varsigma_{12} - \sigma_{33}\tfrac{-1}{w_{82}}\varsigma_{32} - 0 \\
0 &= -(2d_T + \rho)\sigma_{33} - \gamma\sigma_{31}\tfrac{-1}{w_{51}}\gamma\sigma_{13} - \sigma_{32}\tfrac{-1}{w_{81}}\sigma_{23} - (w_{32} + \gamma\varsigma_{13})\tfrac{-1}{w_{52}}\gamma\sigma_{13} - \\
&\quad -\varsigma_{23}\tfrac{-1}{w_{82}}\sigma_{33} - \gamma\sigma_{31}\tfrac{-1}{w_{52}}(\gamma\varsigma_{13} + w_{32}) - \sigma_{33}\tfrac{-1}{w_{82}}\varsigma_{33} - 0
\end{aligned}
$$

---

[4]: In order to avoid some notational problems, the elements of matrix S_1 are denoted as σ_{ij} and the elements of matrix S_2 as ς_{ij}. Likewise, the elements of the vector s_1 are σ_i and the elements of the vector s_2 are ς_i.

Likewise, the system B.15, representing 3 equations, can be written out in the following way:

$$0 = -(\alpha + \rho)\sigma_1 - \gamma\sigma_{11}\frac{-1}{w_{51}}(\gamma\sigma_1 - w_{41}) - \sigma_{12}\frac{-1}{w_{81}}(\sigma_2 - w_{71}) - \gamma\varsigma_{11}\frac{-1}{w_{52}}\gamma\sigma_1 -$$
$$-\varsigma_{21}\frac{-1}{w_{82}}\sigma_3 - \gamma\sigma_{11}\frac{-1}{w_{52}}(\gamma\varsigma_1 - w_{42}) - \sigma_{13}\frac{-1}{w_{82}}(\varsigma_3 - w_{72}) + 0$$

$$0 = -(d_T + \rho)\sigma_2 - (w_{31}\gamma\sigma_{21})\frac{-1}{w_{51}}(\gamma\sigma_1 - w_{41}) - \sigma_{22}\frac{-1}{w_{81}}(\sigma_2 - w_{71})$$
$$-\gamma\varsigma_{12}\frac{-1}{w_{52}}\gamma\sigma_1 - \varsigma_{21}\frac{-1}{w_{82}}\sigma_3 - \gamma\sigma_{21}\frac{-1}{w_{52}}(\gamma\varsigma_1 - w_{42}) - \sigma_{23}\frac{-1}{w_{82}}(\varsigma_3 - w_{72}) + w_1$$

$$0 = -(d_T + \rho)\sigma_3\gamma\sigma_{31}\frac{-1}{w_{51}}(\gamma\sigma_1 - w_{41}) - \sigma_{32}\frac{-1}{w_{81}}(\sigma_2 - w_{71}) - \gamma\varsigma_{13}\frac{-1}{w_{52}}\gamma\sigma_1 -$$
$$-\varsigma_{23}\frac{-1}{w_{82}}\sigma_3 - \gamma\sigma_{31}\frac{-1}{w_{52}}(\gamma\varsigma_1 - w_{42}) - \sigma_{33}\frac{-1}{w_{82}}(\varsigma_3 - w_{72}) + 0$$

Similarly, the twelve equations for player 2 can be written out. This gives 24 non-linear equations with 24 unknowns (σ_{ij}, ς_{ij}, σ_i and ς_i for $i, j = 1..3$). Generally, this does not give unique solutions. However, the sub-game perfect Nash equilibrium must also satisfy the transversality conditions that $\lim_{t\to\infty} \psi_{ij}e^{-\rho t} = 0$ for $i, j = 1, 2$. Following Hanig (1986), it is assumed that with these constraints the system of equations above has a unique solution.

As explained in the main text, using the Riccati coefficients, the controls can be expressed as known functions of the states. This allows a direct use of Pontryagin's maximum principle in the feedback case

B.6 Pareto Case

For the Pareto case, the analysis of the first order conditions and the steady state values is analogous to the open-loop case. The equilibrium values of the control variables in the case of two identical countries is:

$$F = \frac{\frac{w_3}{d_T}(w_1 - (\rho + d_T)w_7) + ((\rho + d_T)w_8 + \frac{w_2}{d_T})\dot{w}_4}{((\rho + d_T)w_8 + \frac{w_2}{d_T})(w_5 + 4\frac{\gamma^2 w_6}{\alpha(\rho+\alpha)}) - \frac{w_3^2}{d_T}} \tag{B.16}$$

$$I = \frac{(w_1 - (\rho + d_T)w_7)(w_5 + 4\frac{\gamma^2 w_6}{\alpha(\rho+\alpha)}) + w_3 w_4}{((\rho + d_T)w_8 + \frac{w_2}{d_T})(w_5 + 4\frac{\gamma^2 w_6}{\alpha(\rho+\alpha)}) - \frac{w_3^2}{d_T}} \tag{B.17}$$

Note that the only difference with Equations B.12 and B.13 is the coefficient 4 in the term $(w_5 + 4\frac{\gamma^2 w_6}{\alpha(\rho+\alpha)})$. In the open-loop case, this coefficient is 2. Likewise, in the situation of N identical players, Pareto optimum has coefficient $2N$, but the open-loop coefficient is still 2. Therefore, with $w_1 > (\rho + d_T)w_7$ and both numerator and denominator positive, as is the case in the numerical example, cooperation gives lower values of F_i and I_i. Knowing that at the steady state, $P = \frac{1}{\alpha}\gamma(F_1 + F_2)$ and $T_i = \frac{1}{d_T}I_i$, it is obvious that these conclusions also hold for P and T_i

Finally, note that for I_i, the difference between the cooperative case and the Nash case disappears if w_3 is put to zero, that is, if the interaction term between F_i and T_i is dropped. In fact the resulting formula for I_i becomes:

$$I = \frac{(w_1 - (\rho + d_T)w_7)}{(\rho + d_T)w_8 + \frac{w_2}{d_T}} \tag{B.18}$$

Therefore, T_i is also unaffected by cooperation as long as $w_3 = 0$.

B.7 Comparison of the Open-loop, Feedback and Cooperative Case

A simple comparison between the open-loop, feedback and cooperative case will be made for the numerical specification of the linear-quadratic case given above with the additional assumption that the interaction between F and T is zero (i.e. $w_3 = 0$). The latter means that the economic subsystem (I and T) is disjunct from the environmental subsystem (P and F).

For the numerical values given, the different control functions for fossil fuel and investment are:

$$
\begin{aligned}
F_i &= 5.470 - 0.0290P \quad \text{joint cooperation case} \\
F_i &= 6.677 - 0.0174P \quad \text{open loop Nash case} \\
F_i &= 7.316 - 0.0157P \quad \text{feedback Nash case}
\end{aligned} \tag{B.19}
$$

$$I_i = 8.065 - 0.0374T \quad \text{all three equilibria}$$

Note that for capital investment, due to $w_3 = 0$, the externality of GHG build-up does not play a role any more, and hence, the control function of investment is the same for the three types of equilibria.

B.8 Renegotiation Proofness

A proof that the strategy profile induced by (π_0, π_1, π_2) is renegotiation proof for small enough values of the discount rate (see Van Damme, 1989, p.211). Given the definition of the present value of pay-offs in Section 4.5, the continuation pay-offs are:

$$PV(\rho; \pi_{j,t}) = \begin{cases} (2,2) \text{ for } j = 0 \text{ and } t = 0 \text{ or for all } j \text{ and } t > 0 \\ (2 - 3\rho, 2 + \rho) \text{ for } j = 1 \text{ and } t = 0 \\ (2 + \rho, 2 - 3\rho) \text{ for } j = 2 \text{ and } t = 0 \end{cases} \qquad (\text{B.20})$$

In order for the continuation pay-offs not to Pareto dominate another (so that deviation from π_0, π_1 or π_2 is not profitable), the following constraints hold:

$$\begin{aligned} 2 &\geq 3\rho + (-1)\rho(1 - \rho) + 2(1 - \rho)^2 \\ 2 - 3\rho &\geq 0\rho + (-1)\rho(1 - \rho) + 2(1 - \rho)^2 \\ 2 + \rho &\geq 2 \end{aligned}$$

This is true for any discount rate lower than 2/3.   *Q.E.D.*

B.9 Specification of the Discrete Time Example

For the example in the text (Table 4.3 – 4.4), the following discrete time version of the model is used[5]:

$$\max_{F_i, I_i} \sum_{t=0}^{\infty} 2T_{it} - 0.005T_{it}^2 + 0F_{it}T_{it} + 10F_{it} - 0.5F_{it}^2 - 0.001P_t^2 - 5I_{it} - 0.5I_{it}^2$$

$$\begin{aligned} s.t. \quad P_{t+1} - P_t &= 0.5F_{1t} + 0.5F_{2t} - 0.005P_t \\ T_{t+1} - T_t &= I_{it} - 0.1T_{it} \end{aligned}$$

Concentrating on fossil fuels[6] and adjusting the continuous time control variables (see Equations B.19) for a discrete time framework, the following approximations of the fossil fuel control variables are used:

$$\begin{aligned} F_{it} &= 5.470 - 0.0290P_{t-1} \quad \text{joint cooperation case} \\ F_{it} &= 6.677 - 0.0174P_{t-1} \quad \text{open loop Nash case} \\ F_{it} &= 7.316 - 0.0157P_{t-1} \quad \text{feedback Nash case} \end{aligned} \qquad (\text{B.21})$$

---

[5]As explained before, the steady state values presented in the main text have been multiplied by a factor 3 to make them comparable with the results in Chapter 3. The steady state of the system given below, have to be multiplied likewise by a factor 3 to make them comparable with the results given in the main text.

[6]This is because players can not influence each others's pay-offs by adjusting their investment control function, due to $w_3 = 0$.

Appendix C

C.1 Description of Feedback Solution

In the feedback Nash case, analogous to the social welfare function of the open-loop case, the following, so called, value function $V(t, P)$ is defined:

$$V(t, P) = \int_t^\infty e^{-\rho t}[\{U[(1 - \tilde{\beta}(\sigma))Y\} + V\{P(\sigma)\}]\,d\sigma \qquad \rho > 0 \qquad \text{(C.1)}$$

where the equation of motion of the GHG concentration is now:

$$\dot{P} = \epsilon(\tilde{\beta})Y + \epsilon^*(\tilde{\beta}^*)Y^* - \delta P \qquad \text{(C.2)}$$

and where $\tilde{\beta}$ is the optimal feedback control.

This value function represents the cost-to-go, given the optimal control. The function $V(t, P)$ has to satisfy the so-called Hamilton/Jacobi/Bellman equation:

$$\rho V(t, P) - \frac{\partial V(t, P)}{\partial t} =$$
$$max_{\beta(t,P)}[U\{(1 - \beta)Y(\sigma)\} + V\{P(\sigma) + \frac{\partial V(t, P)}{\partial P}$$
$$\{\epsilon(\beta)Y + \epsilon^*(\beta^*)Y^* - \delta P\}] \qquad \text{(C.3)}$$

In general, analytical solution for this equation cannot be found. For the linear-quadratic models (or linearised versions of more complex models), the system of Riccati system corresponding to Equation C.3 can be explicitly solved to calculate the controls $\beta(t, P)$ (see Basar & Olsder, 1982, p. 229). The results given in Table 5.1 are calculated using an interative method on the basis of the Riccati equations, as described in Miller & Salmon (1985). For a description, see Chapter 4.

C.2   Proof of Theorems on Linkage

Proof of Theorem 5.3: Without loss of generality, we can state that the one shot game equilibrium is $(C, N; \bar{N}, \bar{C})$.

The following constraints should hold:

$$
\begin{aligned}
(m_1 + \bar{m}_1) &\geq \rho(m_1 + \bar{f}_1) + \rho(1 - \rho)(u_1 + \bar{m}_1) + (1 - \rho)^2(m_1 + \bar{m}_1) \\
(m_2 + \bar{m}_2) &\geq \rho(f_2 + \bar{m}_2) + \rho(1 - \rho)(m_2 + \bar{u}_2) + (1 - \rho)^2(m_2 + \bar{m}_2) \\
\rho(u_1 + \bar{m}_1) + (1 - \rho)(m_1 + \bar{m}_1) &\geq \rho(u_1 + \bar{f}_1) + \rho(1 - \rho)(u_1 + \bar{m}_1) + (1 - \rho)^2(m_1 + \bar{m}_1) \\
\rho(f_2 + \bar{m}_2) + (1 - \rho)(m_2 + \bar{m}_2) &\geq \rho(f_2 + \bar{m}_2) + \rho(1 - \rho)(f_2 + \bar{u}_2) + (1 - \rho)^2(m_2 + \bar{m}_2) \\
\rho(m_1 + \bar{f}_1) + (1 - \rho)(m_1 + \bar{m}_1) &\geq (m_1 + \bar{m}_1) \\
\rho(f_2 + \bar{m}_2) + (1 - \rho)(m_2 + \bar{m}_2) &\geq (m_2 + \bar{m}_2)
\end{aligned}
$$

This is true given our assumptions that:

1. $n_1 + n_2 < u_1 + f_2, u_2 + f_1 < m_1 + m_2$

2. $n_1 + \bar{n}_1, n_2 + \bar{n}_2 < u_1 + \bar{f}_1, \bar{u}_1 + f_1, u_2 + \bar{f}_2, \bar{u}_2 + f_2 < m_1 + \bar{m}_1, m_2 + \bar{m}_2$

Proof of Theorem 5.4.   This proof is similar to the proof above with one-shot equilibrium $(N, N; \bar{N}, \bar{N})$ and with the following constraints:

$$
\begin{aligned}
(m_1 + \bar{m}_1) &\geq \rho(f_1 + \bar{f}_1) + \rho(1 - \rho)(u_1 + \bar{u}_1) + (1 - \rho)^2(m_1 + \bar{m}_1) \\
(m_2 + \bar{m}_2) &\geq \rho(f_2 + \bar{f}_2) + \rho(1 - \rho)(u_2 + \bar{u}_2) + (1 - \rho)^2(m_2 + \bar{m}_2) \\
\rho(u_1 + \bar{m}_1) + (1 - \rho)(m_+\bar{m}) &\geq \rho(n_1 + \bar{n}_1) + \rho(1 - \rho)(u_1 + \bar{u}_1) + (1 - \rho)^2(m_1 + \bar{m}_1) \\
\rho(u_2 + \bar{m}_2) + (1 - \rho)(m_+\bar{m}) &\geq \rho(n_2 + \bar{n}_2) + \rho(1 - \rho)(u_2 + \bar{u}_1) + (1 - \rho)^2(m_2 + \bar{m}_2) \\
\rho(f_1 + \bar{f}_1) + (1 - \rho)(m_1 + \bar{m}_1) &\geq (m_1 + \bar{m}_1) \\
\rho(f_2 + \bar{f}_2) + (1 - \rho)(m_2 + \bar{m}_2) &\geq (m_2 + \bar{m}_2)
\end{aligned}
$$

Literature

Aalbers R.F.T. (1993), "Modeling Environmental Resources", mimeo Tilburg University.

Abreu D., D. Pearce & E. Stachetti (1989), "Renegotiation and Symmetry in Repeated Games", Mimeo, Harvard University.

Alcamo J., M. Amann, J-P. Hettelingh, M Holmberg, L. Hordijk, J. Kämäri, L. Kauppi, P. Kauppi, G, Kornai & A. Mäkelä (1988), "Acidification in Europe: A Simulation Model for Evaluating Control Strategies", IIASA, Laxenburg.

Arrhenius E. & T.W. Waltz (1990), "The Greenhouse Effect: Implications for Economic Development", World Bank Dicussion Papers, No. 78, Washington D.C.

Arrow K.J. (1962), "The Economic Implications of Learning by Doing", *Review of Economic Studies*, **29**, pp. 155-73.

Austin G.P. & W.H. Buiter (1982), "Saddlepoint, a programme for solving continuous time linear rational expectations models", LSE Econometrics Programme, D.P. No.37, November.

Axelrod R. (1984), *The Evolution of Cooperation*, Basic Books, New York, U.S.

Barbier E.B. & A. Markandya (1989), "The Conditions for Achieving Environmentally Sustaiable Development", LEEC Paper 89-01.

Barrett S. (1989), "On the Nature and Significance of International Environmental Agreements", mimeo, London Business School.

Barrett S. (1990), "International Environmental Agreements as Games", Paper presented at the Symposium on "Conflicts and Cooperation in Managing Environmental Resources", November 15-16, 1990, Universitat Gesamthochschule Siegen, Siegen, Germany.

Barrett S. (1991), "The Paradox of International Environmental Agreements", mimeo London Business School.

Barrett S. (OECD) (1991), "Economic Analysis of International Environmental Agreements: Lessons for a Global Warming Treaty", in: OECD, *Responding to Climate Change: Selected Economic Issues*, Paris.

Barro R.J. & X. Sala i Martin (1990), "Public Finance in Models of Economic Growth", NBER Working Paper, No. 3362.

Başar T. & G-J. Olsder (1982), Dynamic Noncooperative Game Theory, Academic Press, London.

Bean C.R. (1990), "Endogenous Growth and the Procyclical Behaviour of Productivity", European Economic Review, 34, pp. 355-63.

Becker G.S., S. Murphy & R. Tamura (1990), "Human Capital, Fertility and Economic Growth", Journal of Political Economy, 98, No. 5.

Becker R.A. (1982), "Intergenerational Equity: The Capital- Environment Trade-Off", Journal of Environmental Economics and Management, 9, pp. 165-185.

Bergman L., H.S.J. Cesar & G. Klaassen (1992), "A Scheme for Sharing the Costs of Reducing Sulphur Emissions in Europe", in: J.J. Krabbe & W.J.M. Heijman (Eds.), National Income and Nature: Externalities, Growth and Steady State, Kluwer Academic Publishers, Dordrecht, the Netherlands.

Bleijenberg A. (1990), "Lastenverschuiving voor Energiebesparing: een case-study naar de gevolgen van een regulerende milieu-heffing", doc\3\7\580 \3AB1 (Centrum voor energiebesparing en schone technologie, Delft).

Boccaccio G. (1349), Decameron, (Engish translation 'Decameron', Pinguin Classics, 1972).

Bohm P. & C.S. Russell (1985), "Comparative analysis of alternative policy instruments", in: A.V. Kneese and J.L. Sweeney (Eds), Handbook of Natural Resource and Energy Economics, Vol. I, pp. 905-26, North-Holland, Amsterdam.

Bohm P. (1990), "Efficiency Aspects of Imperfect Treaties on Global Public Bads: Lessons Learned from the Montreal Protocol", mimeo, World Bank.

Bohm P. (1991), "Incomplete International Cooperation to Reduce CO_2 Emissions: Alternative Policies", Research Papers in Economics, University of Stockholm, 1991:2-WE.

Bolin B. et al. (Eds.) (1986), The Greenhouse Effect, Climate Change, and Ecosystems, SCOPE 29, Wiley & Sons, New York.

Bovenberg A. L. & S. Smulders (1993), "Environmental Quality and Pollution-Saving Technological Change in a Two-Sector Endogenous Growth Model", mimeo Tilburg University.

Brock W A & J A Scheinkman (1976), "Global Asymptotical Stability of Optimal Control Systems with Applications to Dynamic Economic Theory", in: J.D. Pitchford & S.J. Turnovsky (Eds.), Applications to Dynamic Economic Theory, North Holland, Amsterdam.

Brock W.A. (1977), "The Global Asymptotic Stability of Optimal Control: A Survey of Recent Results", in: M.D. Intriligator (Ed.), *Frontiers in Quantitative Economics*, Vol. III-B, North Holland, Amsterdam.

Brown B.J. et al. (1987), "Global Sustainability: Towards Definition", *Environmental Management*, **11**, pp. 713-19.

Brown L.R. et al. (1990), *State of the World 1990*, Worldwatch Institute, Washington D.C.

Carraro C. & D. Siniscalco (1991), "Strategies for the International Protection of the Environment", Paper Presented at the Conference "The Economics of Transnational Commons", 25-27 April, 1991, Siena, Italy.

Cave J. (1987), "Long-Term Compitition in a Dynamic Game: The Cold Fish War", *Rand Journal of Economics*, **18**, pp. 596-610.

Cesar H.S.J. (1989), "Transboundary Pollution and Options for International Abatement Strategies: The Case of Acid Rain in Europe", mimeo Wageningen Agricultural University.

Cesar H.S.J. (1990), "Challenges in Coordinating Abatement of Transfrontier and Global Pollution: A Game Theoretic Approach", Paper presented at the Conference "Economics of the Environment", September 17-19, CentER for Economic Research, Tilburg, The Netherlands.

Cesar H.S.J. (1993-a), "International Cooperation and Technology Transfers in the Case of the Greenhouse Effect", *Structural Change and Economic Dynamics*, **4**, (forthcoming).

Cesar H.S.J. (1993-b), "Modelling Sustainability in a Rudimentary Model of the Greenhouse Effect", Paper presented at the 'ECOZOEK-dag', Tilburg, June 11.

Cesar H.S.J. (1993-c), "Emission Reduction Technology and Human Capital", Paper presented at the 'informal growth seminar', OCFEB, Rotterdam, August 10.

Cesar H.S.J. & P. Della Posta (1991), "Calculating Feedback Solutions with Mark-2", mimeo European University Institute, Florence.

Chandler P. & H. Tulkens (1991), "Strategically Stable Cost Sharing in an Economic-Econogical Negotiation Process", Paper Presented at the Second Annual Conference of the European Association of Environmental and Resource Economists, Stockholm, June 1991.

Clemhout S. & H.Y. Wan (1992), "The Non-uniqueness of Markovian Strategy Equilibrium: The Case of Continuous Time Models for Non-renewable Resources", Paper presented at the Fifth International Symposium on Dynamic Games and Applications, Conches–Grimentz, June 16, 1992, University of Geneva.

Coase R.H. (1960), "The Problem of Social Cost", *Journal of Law and Economics*, **3**, pp. 1-44.

Conches–Grimentz (1992), "Fifth International Symposium on Dynamic Games and Applications", Preprint Volume June 16, 1992, University of Geneva.

Conrad J.M. & C.W. Clark (1987), *Natural Resource Economics*, Cambridge University Press, Cambridge, U.K.

Currie D., G. Holtham & A. Hughes Hallett (1989), "The Theory and Practice of International Policy Coordination: Does Coordination Pay?", CEPR Discussion Paper No. 325.

Dante Alighieri (1309), *La Divina Commedia*, (Engish translation 'The Divine Comedy', Pinguin Classics, 1980).

D' Arge R.C. & K.C. Kogiku (1973), "Economic growth and the Environment", *Review of Economic Studies*, **40**, pp. 61-77.

Dasgupta P. (1982), *Control of Resources*, Basil Blackwell, Oxford.

Dasgupta P. & G.M. Heal (1974), *The Optimal Depletetion of Exhaustible Resources*, Review of Economic Studies.Symposium issue

De Zeeuw A. (1991), "Lectures on Differential Games and Pollution Control", Paper presented at the Autumn Workshop in Environmental Economics, Venice, Sept. 29 –Oct. 5, 1991.

Dutta P.K. (1991), "A Folk Theorem for Stochastic Games", Rochester Center for Economic Research, Working Paper, No. 293.

Enquete-Kommission (1989), *Schutz der Erdatmosphaere*, German Bundestag, Publ. Sect., Bonn; (For citations, the page numbering of the English version "Protecting the Earth's Atmosphere" is taken).

EPA (1990), "Policy Options for Stabilizing Global Climate", Draft Report to Congress, Executive Summary, April 1990.

Farrell J. & E. Maskin (1989), "Renegotiation in Repeated Games", *Games and Economic Behaviour*, **1**, pp. 327-360.

Feichtinger G. & R.F. Hartl (1986), *Optimale Kontrolle Oekonomischer Prozesse*, Walter de Gruyter, Berlin.

Fisher R. (1964), "Fractioning Conflict", in: R. Fisher (Ed.), *International Confilict and Behavioral Science: The Craigville Papers*, New York, Basic Books.

Fershtman C. & M.I. Kamien (1987), "Dynamic Duopolistic Competition with Sticky Prices", *Econometrica*, **55**, pp. 1151-1164.

Folmer H., P. van Mouche & S. Ragland (1991), "Interconnected Games and International Environmental Problems", mimeo Wageningen Agricultural University, The Netherlands.

Forster B.A. (1972), "A Note on the Optimal Control of Pollution", *Journal of Economic Theory*, **5**, pp. 537-539.

Forster B.A. (1975), "Optimal Pollution Control with a Nonconstant Exponential Rate of Decay", *Journal of Environmental Economics and Management*, **2**, pp. 1-6.

Forster B.A. (1977), "On a One State Variable Optimal Control Problem: Consumption Pollution Trade-Offs", in: J.D. Pitchford & S.J. Turnovsky (Eds.), *Applications of Control Theory to Economic Analysis*, North Holland, Amsterdam.

Forster B.A. (1980), "Optimal Energy Use in a Polluted Environment", *Journal of Environmental Economics and Management*, **7**, 321-333.

Friedman A. (1971), *Differential Games*, Wiley-Interscience, New York.

Friedman J. (1971), "A Noncooperative Equilibrium for Supergames", *Review of Economic Studies*, **38**, pp. 1-12.

Friedman J. (1986), *Game Theory with Applications to Economics*, Oxford University Press, Oxford.

Fudenberg D. & J. Tirole (1991), *Game Theory*, the MIT Press, Cambridge Massachusetts.

Fujii Y. (1990), "An Assessment of the Responsibility for the Increase in the CO_2 Concentration and Inter-generational Carbon Accounts", Working Paper, IIASA, WP-90-55.

Gottinger H.W. (1990), "Economic Models of Optimal Energy Use inder Global Environmental Constraints", Working Paper, EV2, Oxford Institute for Energy Studies, Oxford.

Gottinger H.W. (1991), "Policy Models of Long-run Growth under Global Environmental Constraints", Working Paper, EV3, Oxford Institute for Energy Studies, Oxford.

Gradus H.J.M. & P.M. Kort (1991), "Optimal Taxation on Profit and Pollution with a Macroeconomic Framework", Reseach Memorandum, Tilburg University, FEW 484.

H.J.M. Gradus & S. Smulders (1993), "The Trade-Off Between Environmental Care and Long-Term Growth: Pollution in Three Prototype Growth Models", *Journal of Economics*, **58**, (Forthcoming).

Grubb M. (1989), *The Greenhouse Effect: Negotiating Targets*, The Royal Institute of International Affairs, London.

Grubb M. & J.K. Sebenius (1991), "Participation, Allocation and Adaptability in International Tradeable Emission Permit Systems for Greenhouse Gas Control", Paper presented at the OECD workshop on tradeable emission permits to reduce greenhouse gases.

Gruver G.W. (1976), "Optimal Investment in Pollution Control Capital in a Neoclassical Growth Context", *Journal of Environmental Economics and Management,* **3,** pp. 165-177.

Hanig M. (1986), "Differential Gaming Models of Oligopoly", MIT (PhD dissertation).

Hanley N. (1992), "Are There Environmental Limits to Cost Benefit Analysis?", *Environmental and Resource Economics,* **2,** pp. 33-59.

Hardin G. (1968), "The Tragedy of the Commons", *Science,* **162,** .

Harrod R.F. (1948), *Towards a Dynamic Economy,* St. Martin's Press, London.

Heal G.M. (1991), "International Negotiations on Emission Control", mimeo Graduate School of Business, Colombia University.

Hoel M. (1989), "Global Environmental Problems: The Effects of Unilateral Actions Taken by One Country", mimeo University of Oslo.

Hoel M. (1990), "Emission Taxes in a Dynamic Game of CO2 Emissions", Working Paper, Nr. 7, University of Oslo.

Hope C. (1991), "PAGE", mimeo Cambridge University, Cambridge, U.K..

Hotelling H. (1931), "The economics of exhaustible resources", *Journal of Political Economy,* **39,** 137-175.

Hung V.T.Y., P. Chang & K. Blackburn (1992), "Endogenous Growth, Environment and R & D", Paper presented at the Workshop "The International Dimension of Environmental Policy" in Milan, 22-24 October 1992.

IPCC (1990), "IPCC First Assessment Report: Overview, WGI Policymakers Summary, WGII Policymakers Summary, WGIII Policymakers Summary", Oxford University Press.

Johanson L. (1959), "Substitution vs. Fixed Production Coefficients in the Theory of Economic Growth", *Econometrica,* **27,** pp. 157-76.

Jorgenson D.W. & Wilcoxen, P.J. (1990), "Reducing US carbon dioxide emissions: the cost of different goals", Harvard Institute of Economic Research, Discussion Paper, nr. 1575.

Kaitala V. & M. Pohjola (1988), "Optimal Recovery of a Shared Resource Stock: A Differential Game Model with Efficient Memory Equilibria", *Natural Resource Modeling,* **3,** pp. 91-119.

Kaitala V. &. M. Pohjola (1990), "Economic Development and Agreeable Redistribution in Capitalism: Efficient Game Equilibria in a Two-Class Neo-Classical Growth Model", *International Economic Review,* **32,** pp. 421-438.

Kaitala V., M. Pohjola & O. Tahvonen (1990), "Transboundary Air Pollution and Soil Acidification; A Dynamic Analysis of an Acid Rain Game Between Finland and the USSR", ETLA, Working Paper, No. 344, Helsinki.

Keeler E., M. Spence & R. Zeckhauser (1971), "The Optimal Control of Pollution", *Journal of Economic Theory,* **4,** pp. 19-34.

Keohane R.O. (1984), *After Hegenomy: Cooperation and Discord in the World Political Economy,* Princeton University Press, Princeton, U.S.

Keohane R.O. (1986), "Reciprocity in International Relations", *International Organization,* **40,** 1, Winter 1986.

King R.G. & S. Rebelo (1990), "Public Policy and Economic Growth: Developing Neoclassical Implications", NBER Working Paper, No. 3338.

Klaassen G. & H.M.A. Jansen (1989), "Economic Principles for Allocating the Costs of Reducing Sulphur Emissions in Europe", Working Paper, Institute for Environmental Studies, Amsterdam.

Kolstad C.D. (1992), "Looking vs. leaping: information acquisition vs. emission control in global warming policy", mimeo, Institute for Environmental Studies and Department of Economics, University of Illinois.

Leggett J. (1990), "The Science of Climate-Modelling and a Perspective on the global-Warming Debate", in: J. Leggett (Ed.), *Global Warming, The Greenpeace Report,* Oxford University Press, Oxford.

Levhari D. & L.J. Mirman (1980), "The great fish war: an example using a dynamic Cournot-Nash solution", *Bell Journal of Economics,* **11,** 322-344.

Lockwood B. (1990), "The Folk Theorem in Stochastic Games with and without Discounting", mimeo, University of London.

Lucas R.E. Jr. (1988), "On the Mechanics of Economic Development", *Journal of Monetary Economics,* **22,** pp. 3-42.

Luptáčik M. & U. Schubert (1982), "Optimal Investment Policy in Productive Capacity and Pollution Abatement Processes in a Growing Economy", in: G. Feichtinger (Ed.), *Optimal Control Theory and Economic Analysis,* North Holland, Amsterdam.

Machiavelli N. (1513), *Il Principe,* (Engish translation 'The Prince', Pinguin Classics, 1988).

Mäler K-G. (1974), Environmental Economics, A Theoretical Inquiry, Baltimore.

Mäler K-G. (1989), "The Acid Rain Game", in: H. Folmer & E. van Ierland (Eds.), *Valuation Methods and Policy Making in Environmental Economics,* Elsevier Science Publishers B.V. Amsterdam.

Mäler K-G. (1991), "The Acid Rain Game II", Paper presented at the Autumn Workshop in Environmental Economics, Venice, Sept. 29 –Oct. 5, 1991.

Manne, A.S. & R.G. Richels (1990), "Global CO_2 Emission Reductions – the Impacts of Rising Energy Costs", Electric Power Research Institute, Poalo Alto, U.S.A.

Markink A.J. & F. van der Ploeg (1991), "Dynamic Polic in Linear Models with Rational Expectations of Future Events: A Computer Package", *Computer Science in Economics and Management,* **4,** pp. 175-199.

Martin L.L. (1992-a), "Interests, Power and Multilateralism", *International Organization,* **46,** Fall 1992.

Martin L.L. (1992-b), Coercive Cooperation: Explaining Multilateral Economic Sanctions, Princeton University Press, Princeton, U.S.

McGinnis M.D. (1986), "Issue Linkage and the Evolution of International Cooperation", *Journal of Conflict Resolution,* **30,** pp.141-70.

Meadows D.H. et al. (1972), The Limits to Growth, Universe Books, New York.

Miller M. & M. Salmon (1985), "Policy coordination and dynamic games", in: W.H. Buiter & R.C. Marston (Eds.), *International economic policy coordination,* Cambridge University Press, Cambridge.

Morgenstern (1991), "Towards a Comprehensive Approach to Global Climate Change Mitigation", *AER – Papers & Proceedings,* **81,** pp. 140-145.

Musu I. (1990), "Optimal Accumulation and Control of Environmental Quality", *Rivista Internationale di Scienze Economiche e Commerciali,* **37,** pp. 193-202.

Musu I. (1991), "The Interdependence between Environment and Development: Marine Pollution in the Mediterranean Sea", Paper Presented at the Conference "The Economics of Transnational Commons", 25-27 April, 1991, Siena, Italy.

Musu I. (1992), "Sustainable economy, technology and time preference", paper presented at the third annual meeting of the EAERE, Krakow 1992.

NAS 1991, Policy Implications of Global Warming: the Mitigation Panel, Washington.

Nerlove M. & A. Meyer (1991), "Endogenous Fertility and the Environment: The Parable of Firewood", Paper Presented at the Second Annual Conference of the European Association of Environmental and Resource Economists, Stockholm, June 1991.

Nitze W.A. (1990), The Greenhouse Effect: Formulating a Convention, The Royal Institute of International Affairs, London.

Nordhaus W. (1982), "How Fast Should We Graze the Global Commons", *American Economic Review,* **72,** Nr. 2, pp. 242-246.

Nordhaus W. (1989), "The Economics of the Greenhouse Effect", Paper prepared for the 1989 Meetings of the International Energy Workshop and the MIT Symposium on Environment and Energy, August 1989.

Nordhaus W. (1990), "To Slow or not to Slow: The Economics of the Greenhouse Effect", Cowles Foundation Discussion Paper.

Nordhaus 1991, "A Sketch of the Economics of the Greenhouse Effect", *AER – Papers & Proceedings,* **81,** pp. 146-150.

OECD (1991), "Greenhouse Gas Emissions: The Energy Dimension", (together with International Energy Agency) Paris.

OECD (1992), "New Issues, New Results: The OECD's Second Survey of the Macroeconomic Costs of Reducing CO_2 Emissions", Economics Department Working Paper No. 123 (by P. Hoeller, A. Dean & M. Hayafuji).

Papavassilopoulos, G.P. & J.B. Cruz (1979), "Nonclassical control problems and Stackelberg games", *IEEE Trans. Aut. Contr.,* **AC-24,** 155-165.

Parson E. A. (1991), "Protecting The Ozone Layer: The Evolution and Impact of International Institutions", International Environmental Institutions Project, March 8, 1992, Harvard University.

Parson, E.A. & R.J. Zeckhauser (1992), "The Tragedy of the Unbalanced Commons: Climate Change and Other Asymmetric International Problems of Collective Action", Paper presented at the SCCN Conference, Stanford, Febr. 14-15 1992.

Pearce D.W. & R.K. Turner (1990), Economics of Natural Resources and the Environment, Harvester Wheatsheaf, London, U.K.

Pethig R. (1990), "Optimal Pollution Control, Irriversibilities, and the Value of Future Information", Discussion Paper, No. 6-90, University of Siegen.

Pezzey J. (1989), "Economic Analysis of Sustainable Growth and Sustainable Development", Environment Department Working Paper, No. 15, World Bank.

Plourde C.G. (1972), "A Model of Waste Accumulation and Disposal", *Canadian Journal of Economics*, **5**, pp. 119-125.

Raiffa, H. (1982), *The Art and Science of Negotiation*, Harvard University Press, Cambridge, Massachusetts, U.S.

Ramsey F.P. (1928), "A Mathematical Theory of Saving", *Economic Journal*, **38**, pp. 543-59.

Reinganum & Stokey 1985, "Oligopoly Extraction of a Common Property Resource: The Importance of the Period of Commitment in Dynamic Games", *International Economic Review*, **26**, pp. 161-173.

Romer P.M. (1986), "Increasing Returns and Long-Run Growth", *Journal of Political Economy*, **94**, pp. 1002-1037.

Romer P.M. (1990), "Endogenous Technological Change", *Journal of Political Economy*, **98**, pp. 71-102.

Sebenius, J.K. (1983), "Negotiation Arithmatic: Adding and Subtracting Issues and Parties", *International Organization*, **37**, No.1.

Sebenius, J.K. (1984), *Negotiating the Law of the Sea: Lessons in the Art and Science of Reaching Agreement*, Harvard University Press, Cambridge, Massachusetts, U.S.

Siebert H. (1992), *Economics of the Environment: Theory and Policy*, Third revised edition, Springer Verlag, Berlin.

Smith V.L. (1972), "Dynamics of Waste Accumulation: Disposal versus Recycling", *Quarterly Journal of Economics*, **86**, pp. 600-616.

Smith V.L. (1977), "Control Theory Applied to Natural and Environmental Resources: An Exposition", *Journal of Environmental Economics and Management*, **4**, 1-24.

Snidal D. (1985), "Coordination versus Prisoners' Dilemma: Implications for International Cooperation and Regimes", *American Political Science Review*, **79**, pp.923-42.

Solow R.M. (1956), "A Contribution to the Theory of Economic Growth", *Quarterly Journal of Economics*, **70**, pp. 65-94.

Solow R.M. (1973), "Is the End of the World at Hand", *Challenge*, , March/April, pp. 39-50.

Solow R.M. (1974), "The Economics of Resources or the Resources of Economics", *Scandinavian Journal of Economics*, **88**, pp. 141-156.

Starrett D. (1972), "Fundamental Non-Convexities in the Theory of Externalities", *Journal of Economic Theory*, **4**, pp. 180-199.

Stein A. (1980), "The Politics of Linkage", *World Politics,* **32,** pp. 62-81.

Stewart 1990, "Technology Transfer for Development", in: R.E. Everson & G. Ramis (Eds.), *Science and Technology; Lessons for Development Policy,* Intermediate Technolgy Publications, Westview Press, London.

Stiglitz J. (1974), "Growth with Exhaustible Natural Resources: Efficient and Optimal Growth Paths", *Review of Economic Studies,* **xx,** Symposium Issue.

Tahvonen O. & J. Kuuluvainen (1990), "The Existance of Steady States in Growth Models with Renewable Resources and Pollution", Discussion Papers, University of Helsinki, No. 299, 8/10/90.

Tahvonen O., H. von Storch & J. Xu (1992), "Optimal Control of CO_2 Emissions", Paper Presented at the Annual Conference of the European Association of Environmental and Resource Economists, Cracow, Poland, June 16-19, 1992.

Tietenberg T. (1985), Emission Trading, Resources for the Future, Washington D.C.

Tollison R.D. & T.D. Willett (1979), "An Economic Theory of Mutually Advantageous Issue Linkage in International Negotiations", *International Organization,* **33,** pp. 425-49.

Tolwinski B. 1992, "Assessing the Impact of International Environmental Policies: A Dynamic Game Perspective", Paper presented at the Fifth International Symposium on Dynamic Games and Applications, Conches–Grimentz, June 16, 1992, University of Geneva.

Tolwinksi B., A. Haurie & G. Leitman 1986, "Cooperative Equilibria in Differential Games", *Journal of Mathematical Analysis and Applications,* **119,** pp. 182-202.

Tsutsui S. & K. Mino (1990), "Nonlinear Strategies in Dynamic Duopolistic Competition with Sticky Prices", *Journal of Economic Theory,* **52,** pp. 136-161.

Tversky A., P. Slovic & D. Kahneman (1990), "The Causes of Preference Reversal", *American Economic Review,* **80,** No. 1.

Ullman R.H. 1983, "Redefining Security", *International Security,* **8,** 129- 153.

Ulph A. (1992), "The Choice of Environmental Policy Instruments and Strategic International Trade", in: R. Pethig (Ed.), *Conflicts and Cooperation in Managing Environmental Resources,* Springer, Berlin.

Ulph D. (1992), "Strategic Innovation and Strategic Environmental Policy", Paper presented at the Workshop "The International Dimension of Environmental Policy" in Milan, 22-24 October 1992.

UNCED (1991),, Global Climate Change: The Role of Technology Transfer, A Report for the United Nations Conference on Environment and Development.

Van Damme E. (1989), "Renegotiation Proof Equilibria in Repeated Prisoner's Dilemma", *Journal of Economic Theory*, **47**, p.206-217.

Van de Klundert T. (1990), "The Ultimate Consequences of the New Growth Theroy; An Introduction to the Views of M. Fitzgerald Scott", CentER for Economic Research Discussion Paper, No. 9054, Tilburg, The Netherlands.

Van der Ploeg F. & A.J. De Zeeuw (1991), "A Differential Game of International Pollution Control", *System & Control Letters*, **17**, pp. 409-414.

Van der Ploeg F. & A.J. De Zeeuw (1992), "International Aspects of Pollution Control", *Environmental & Resource Economics*, **3**, pp. 117-139.

Van der Ploeg F. & C. Withagen (1991), "Pollution Control and the Ramsey Problem", *Environmental and Resource Economics*, **1**, pp. 215-236.

Van der Ploeg F. & Ligthart (1993), "Sustainable Growth and Renewable Resources in the Global Economy", mimeo University of Amsterdam.

Van Long N. (1984), "Risk and Resource Economics: The State of the Art", in: D.W. Pearce, H. Pearce & I. Walter (Eds.), *Risk and the Political Economy of Resource Development*, MacMillan Press, London.

WCED (1987), *Our Common Future*, (The Brundtland Report), New York.

Wirl F. (1991), "Complex Dynamic Environmental Policies", mimeo Institute of Energy Economics, Technical University of Vienna.

Wymer C. (1992), "Computer Programs: APREDIC", mimeo London School of Economics.

Xepapadeas A.P. (1991), "Global Environmental Problems and 'Induced' Technical Change: The Greenhouse Effect", Paper Presented at the Second Annual Conference of the European Association of Environmental and Resource Economists, Stockholm, June 1991.

Yeung D.W.K., & M.T. Cheung (1992), "Capital Accumulation Subject to Pollution Control: a Differential Game with a Feedback Nash Equilibrium", Paper presented at the Fifth International Symposium on Dynamic Games and Applications, Conches–Grimentz, June 16, 1992, University of Geneva.

Young, H.P. (1991), "Sharing the Burden of Global Warming", mimeo School of Public Affairs, University of Maryland.

Zylicz T. (1992), "Implementing Environmental Policies in Central & Eastern Europe", Paper presented at the Second Meeting of the International Society for Ecological Economics.

List of Figures

List of Tables

Vol. 373: A. Billot, Economic Theory of Fuzzy Equilibria. XIII, 164 pages. 1992.

Vol. 374: G. Pflug, U. Dieter (Eds.), Simulation and Optimization. Proceedings, 1990. X, 162 pages. 1992.

Vol. 375: S.-J. Chen, Ch.-L. Hwang, Fuzzy Multiple Attribute Decision Making. XII, 536 pages. 1992.

Vol. 376: K.-H. Jöckel, G. Rothe, W. Sendler (Eds.), Bootstrapping and Related Techniques. Proceedings, 1990. VIII, 247 pages. 1992.

Vol. 377: A. Villar, Operator Theorems with Applications to Distributive Problems and Equilibrium Models. XVI, 160 pages. 1992.

Vol. 378: W. Krabs, J. Zowe (Eds.), Modern Methods of Optimization. Proceedings, 1990. VIII, 348 pages. 1992.

Vol. 379: K. Marti (Ed.), Stochastic Optimization. Proceedings, 1990. VII, 182 pages. 1992.

Vol. 380: J. Odelstad, Invariance and Structural Dependence. XII, 245 pages. 1992.

Vol. 381: C. Giannini, Topics in Structural VAR Econometrics. XI, 131 pages. 1992.

Vol. 382: W. Oettli, D. Pallaschke (Eds.), Advances in Optimization. Proceedings, 1991. X, 527 pages. 1992.

Vol. 383: J. Vartiainen, Capital Accumulation in a Corporatist Economy. VII, 177 pages. 1992.

Vol. 384: A. Martina, Lectures on the Economic Theory of Taxation. XII, 313 pages. 1992.

Vol. 385: J. Gardeazabal, M. Regúlez, The Monetary Model of Exchange Rates and Cointegration. X, 194 pages. 1992.

Vol. 386: M. Desrochers, J.-M. Rousseau (Eds.), Computer-Aided Transit Scheduling. Proceedings, 1990. XIII, 432 pages. 1992.

Vol. 387: W. Gaertner, M. Klemisch-Ahlert, Social Choice and Bargaining Perspectives on Distributive Justice. VIII, 131 pages. 1992.

Vol. 388: D. Bartmann, M. J. Beckmann, Inventory Control. XV, 252 pages. 1992.

Vol. 389: B. Dutta, D. Mookherjee, T. Parthasarathy, T. Raghavan, D. Ray, S. Tijs (Eds.), Game Theory and Economic Applications. Proceedings, 1990. IX, 454 pages. 1992.

Vol. 390: G. Sorger, Minimum Impatience Theorem for Recursive Economic Models. X, 162 pages. 1992.

Vol. 391: C. Keser, Experimental Duopoly Markets with Demand Inertia. X, 150 pages. 1992.

Vol. 392: K. Frauendorfer, Stochastic Two-Stage Programming. VIII, 228 pages. 1992.

Vol. 393: B. Lucke, Price Stabilization on World Agricultural Markets. XI, 274 pages. 1992.

Vol. 394: Y.-J. Lai, C.-L. Hwang, Fuzzy Mathematical Programming. XIII, 301 pages. 1992.

Vol. 395: G. Haag, U. Mueller, K. G. Troitzsch (Eds.), Economic Evolution and Demographic Change. XVI, 409 pages. 1992.

Vol. 396: R. V. V. Vidal (Ed.), Applied Simulated Annealing. VIII, 358 pages. 1992.

Vol. 397: J. Wessels, A. P. Wierzbicki (Eds.), User-Oriented Methodology and Techniques of Decision Analysis and Support. Proceedings, 1991. XII, 295 pages. 1993.

Vol. 398: J.-P. Urbain, Exogeneity in Error Correction Models. XI, 189 pages. 1993.

Vol. 399: F. Gori, L. Geronazzo, M. Galeotti (Eds.), Nonlinear Dynamics in Economics and Social Sciences. Proceedings, 1991. VIII, 367 pages. 1993.

Vol. 400: H. Tanizaki, Nonlinear Filters. XII, 203 pages. 1993.

Vol. 401: K. Mosler, M. Scarsini, Stochastic Orders and Applications. V, 379 pages. 1993.

Vol. 402: A. van den Elzen, Adjustment Processes for Exchange Economies and Noncooperative Games. VII, 146 pages. 1993.

Vol. 403: G. Brennscheidt, Predictive Behavior. VI, 227 pages. 1993.

Vol. 404: Y.-J. Lai, Ch.-L. Hwang, Fuzzy Multiple Objective Decision Making. XIV, 475 pages. 1994.

Vol. 405: S. Komlósi, T. Rapcsák, S. Schaible (Eds.), Generalized Convexity. Proceedings, 1992. VIII, 404 pages. 1994.

Vol. 406: N. M. Hung, N. V. Quyen, Dynamic Timing Decisions Under Uncertainty. X, 194 pages. 1994.

Vol. 407: M. Ooms, Empirical Vector Autoregressive Modeling. XIII, 380 pages. 1994.

Vol. 408: K. Haase, Lotsizing and Scheduling for Production Planning. VIII, 118 pages. 1994.

Vol. 409: A. Sprecher, Resource-Constrained Project Scheduling. XII, 142 pages. 1994.

Vol. 410: R. Winkelmann, Count Data Models. XI, 213 pages. 1994.

Vol. 411: S. Dauzère-Péres, J.-B. Lasserre, An Integrated Approach in Production Planning and Scheduling. XVI, 137 pages. 1994.

Vol. 412: B. Kuon, Two-Person Bargaining Experiments with Incomplete Information. IX, 293 pages. 1994.

Vol. 413: R. Fiorito (Ed.), Inventory, Business Cycles and Monetary Transmission. VI, 287 pages. 1994.

Vol. 414: Y. Crama, A. Oerlemans, F. Spieksma, Production Planning in Automated Manufacturing. X, 210 pages. 1994.

Vol. 415: P. C. Nicola, Imperfect General Equilibrium. XI, 167 pages. 1994.

Vol. 416: H. S. J. Cesar, Control and Game Models of the Greenhouse Effect. XI, 225 pages. 1994.